TWR and Jaguar's XJS

Inside Tom Walkinshaw's Group A Racing Team

TWR and Jaguar's XJS

Inside Tom Walkinshaw's Group A Racing Team

ALLAN SCOTT

First published December 2012

ISBN: 978-0-9876665-5-0

Published by
John Douglas Publishing
P.O. Box 14-022
Christchurch Airport, New Zealand

Typeset in Minion 12/16

Contents

Abbreviations and measurements

EngDiv.	Race Engine Division
BSCC	British Saloon Car Championship
ETCC	European Touring Car Championship
mph	miles per hour
bhp	brake horse power
rpm	revolutions per minute
BMEP	Break Mean Effective Pressure
lb/ft	pounds/feet
lb/in	pounds per inch
in	inch
psi	pounds per square inch
RON	research octane number
MON	motor octane number
TDC	top dead centre
ATDC	after top dead centre
BTDC	before top dead centre
BBDC	before bottom dead centre
ABDC	after bottom dead centre
1 inch	= 2.54 millimetres
1 mile	= 1.61 kilometres
2.24 lbs	= 1 kilogram
1.34 bhp	= 1 kilowatt
1 lb/ft	= 1.345 Newtons

chapter one

Evaluating the prospect of working with Jaguar

In late 1981 Tom Walkinshaw Racing (TWR) was involved in an evaluation programme for Jaguar Cars after secret meetings with their directors at the company's Browns Lane headquarters in Coventry. These meetings had centred on the proposal put forward by team owner Tom Walkinshaw to run a team of Jaguar XJSs the following year in the European Touring Car Championship (ETCC) to the Group A regulations. These regulations would replace the previous Fédération Internationale de l'Automobile (FIA) Group 2 rules.

TWR had emerged as the top saloon car racing team in Britain by this time operating out of one factory building on the Kidlington Industrial Estate a few miles from Oxford, just over an hour's drive due west of London. Tom Walkinshaw had decided from the early stages of his establishment of TWR at Kidlington that this would be an expanding and diversified company. TWR were the UK distributors for the Alpina range of performance products for BMW vehicles, as well as Dunlop racing tyres via DART.

Earlier successes had come from an association with Mazda Cars GB to produce a very successful three year programme in the British Saloon Car Championship [BSCC] with Mazda's RX-7s from 1979 to 1981, culminating in winning the Championship in both 1980 and 1981. This had led to an after market product range being designed and manufactured for Mazda vehicles, as well as opening a Mazda dealership at Dorchester in Oxford.

I had joined TWR in late 1979 after arriving from New Zealand in May of that year, and had begun working for TWR's engine builders, Racing Services [Engines] in Twickenham, London. During that year I had continued to develop and rebuild the team's

The Spa 24 Hours winning Mazda RX-7.

rotary engines for the single RX-7 race car and at year's end Tom had invited me to set up a new department at TWR's Kidlington base, to be called the Race Engine shop or EngDiv. for short. EngDiv.'s task was to develop the various engine programmes in support of the now two TWR Mazda teams running in the British and Belgium Championships in 1980. This had culminated in winning the prestigious Spa 24 Hours event in 1981, the first outright win in a FIA 24 hour endurance race for Mazda's rotary engine, and is covered in my book on the early days with TWR and Mazda. Towards the end of 1980 I had hired Barry Holland which doubled the size of EngDiv. to help with the growing work load.

In 1981 the addition of the British Leyland works Rover team had more than doubled the size of TWR over night. Importantly it was this Rover programme that had begun TWR's relationship with the British automotive industry and showed the benefits of working with similar focused outcomes and cultures that had not always been available when working with extended communication links to Japan with Mazda. The TWR Rover programme in the BSCC was now at the end of its first year and would be the start of a very successful six year relationship in motorsport with the Rover SD1 and later Vitesse model. This story is covered in my book on TWR's Rover racing programme from 1980 to 1986.

By the end of 1981 it was time for Tom Walkinshaw [TW] to take some major decisions on future initiatives and philosophy for his rapidly expanding businesses associated with motor racing. It was after the successful Rover V8 engine development programme

TWR's Rover team at Oulton Park in 1981.

in 1981 that TW had decided we could perhaps do the same with Jaguar's V12 engine, although an earlier racing programme using Jaguar's XJC in 1976/77 by the Broadspeed team had produced less than satisfactory results, and had been cancelled amidst general disappointment for those involved.

The approach to Jaguar Cars in Coventry was the second part of a new philosophy for his company: to become the leading British car racing team promoting British automotive manufacturers against their European and Japanese opposition, on the racing circuits of Europe.

It was decided to do a preliminary investigation of Jaguar and this included some of its earlier racing history as well as their current situation, which alarmingly appeared to produce some doubts as to its future viability.

In April 1980, Jaguar had separated from Rover and Triumph within the fractured British Leyland conglomerate, and once again had gained a greater degree of engineering independence in its bid to survive and pursue its niche markets which had been virtually destroyed under the British Government-supported Leyland umbrella. Earlier success in motor racing during the 1950s had launched the then tiny British Jaguar marque on to the international scene, allowing it to gain vital additional export sales in the American car market. Jaguar's Competitions Department at Browns Lane had built and prepared their own C- and D-Type sports racing cars from 1951 to 1957, scoring five Le Mans victories, and in the process had introduced the concept of a high-performance family saloon car with a distinctive racing pedigree. Jaguar therefore had a strong racing heritage, although it came from a distant era and had lost most of its effect on modern customers.

Jaguar had also suffered since losing the drive and charisma of its founder Sir William Lyons on his retirement in 1971. His replacement, former Competitions Department Chief Lofty England, though a dedicated Jaguar man, had been unable to cope with the British Leyland engineering interference and had left in 1974. Despite all this, however, Jaguar was still surviving while most of the other famous British marques, including Triumph, Standard, Wolseley, Riley, and many others had disappeared into oblivion.

After Jaguar Cars had come within months of closure in late 1979, South African-born Michael Edwardes had been brought in by the British Government; he had to save Jaguar's parent company British Leyland – or close it down. Previously the head of Union Carbide, Edwardes had then laid down the law to the unions, salvaged what he could from the fractured enterprise, and was later knighted for his efforts. He had immediately employed 40-year-old former Unipart and then Ferguson Tractors executive, John Egan, and put him into the Browns Lane headquarters to determine whether Jaguar was even worth saving. In this Egan was helped by existing employees Jim Randle and Roger Putnam, heading up the engineering and marketing departments respectively.

Soon after his arrival, Egan had been impressed with a visit by Jaguar enthusiasts from Australia, led by Les and Bronwyn Hughes, publishers of *Jaguar Magazine* in that country. He reasoned that if Jaguar owners could travel from the other side of the world to stand on what was for them hallowed ground, then there was considerable depth and goodwill still remaining with the Jaguar brand.

Part of Jaguar's point of difference was its large alloy V12 engine in its main market, the USA, where cast-iron pushrod V8s were the norm. But Jaguar sales had reached a low point and Egan's team had to make the difficult decision to cease production of the V12 XJS Coupe in 1980. While the V12 engined saloons continued to be made they too were selling in much reduced numbers as the market became more aware of increasing fuel prices and the USA 'Gas Guzzler' tax imposed on this engine. It was the older 6-cylinder XK model alone that propped up sales during this period and serious consideration was given to halting V12 production entirely. The XK engine's replacement, however, the new AJ6 engine, used much of the same production facility and machinery as the V12 and this persuaded the management to keep the V12 line operational. It was hoped that a new upgraded version of the V12 with its May fireball cylinder head might eventually make a difference to sales, but even with that upgrade to the V12 engine, its performance and efficiency was disappointing in full USA emission specification.

Re-establishing Jaguar's primacy in the use of high-performance single overhead camshaft V12 engines would need a huge marketing effort if there was ever a chance to successfully promote the British marque's only real point of difference in all its markets.

By early 1981 Jaguar Cars was re-emerging from the loss of its unique status as a specialist car builder and needed to become profitable in order to again be an independent entity (an aim that was eventually achieved in 1984).

Would a return to motor racing now be an option for John Egan to consider in his quest to revitalise the Jaguar brand? The earlier XK 6-cylinder engine had contributed to Jaguar's emergence through its many successes at motor racing's premier endurance event, the Le Mans 24 hour. Could this success be repeated on European race tracks all these years later, and produce similar results promoting the V12 engine? This was what Tom Walkinshaw was thinking with his very timely approach in 1981, despite Jaguar's uncertain future.

Late in 1981 Egan had then put forward some new ideas for an urgent marketing programme to re-establish Jaguar's former niche identity, and part of that new focus could include the possibility of a motor-racing programme with TWR, using Jaguar's V12-engined XJS Coupe.

The first official meeting to put TWR's initial proposal to Jaguar got off to a slow start. The four of us aboard the helicopter to Browns Lane had to turn around in thick fog and return to Kidlington, and another meeting had to be hastily convened.

John Egan's first question at this next meeting was to ask why the earlier Broadspeed XJC programme had failed – a question we then set out to answer during the last three months of 1981. His second statement was to the effect that Jaguar had no money! This statement was disappointing but did not seem to have the negative effect I had expected on TW.

Access to remote brake bleed nipples was from underneath the race car, adding to servicing complexity.

direction under braking, a recipe for fatigue of these items under hard racing conditions.

The Jaguar four-speed gearbox was being pushed beyond its design limits, and this was a major weakness even though the gear designs and ratios could be changed. The unit fitted to the Broadspeed XJC V12 was an earlier close-ratio E-Type gearbox designed for a 280bhp engine. It was severely stressed because the power level from the racing 5400cc V12 engine was now 450-480bhp. Some other transmission components had failed, but these were teething issues which, in time, could have been resolved.

The position of the engine's tubular exhaust hampered gearbox and clutch repairs, leading to unacceptable down times at race meetings. Heat transfer to the gearbox was another problem with the close proximity of the exhaust system. We concluded that servicing the race car at a race meeting was compromised by its complexity.

On the positive side, the XJC race car had some excellent features. The engine produced much higher power outputs than its BMW opposition, though the V12's lubrication system let this positive aspect down and would be an important issue that TWR/EngDiv. had to address.

The wheels were huge, and fitted with ultra-low-profile racing slick tyres which were positively located on the wheel rims by four fixing bolts. The front tyres were 300/650 on rims 19 inches in diameter and 13 inches wide. The rears were 330/700 on 19 inch

BELOW
Left: Tubular exhausts transferred heat to the gearbox. Quick gearbox changes were compromised.
Right: Huge rear wheels – transmission coolers were fitted on either side at the rear.

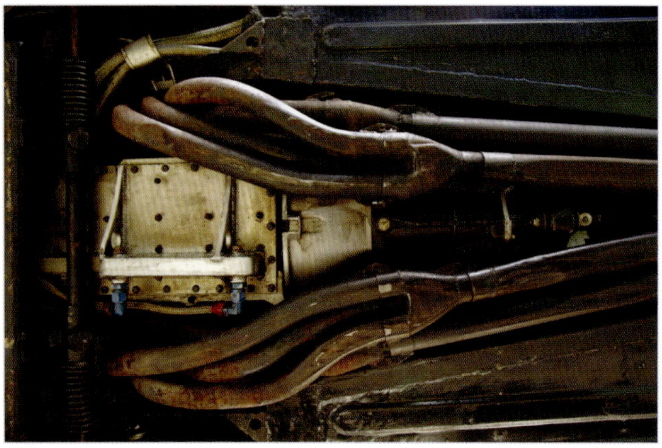

diameter rims 14.5 inches wide. Such huge tyres were more than enough to control the 1700kg weight of the car, and gave good balance and handling.

Aerodynamic down force was produced from the rear body work.

Oil coolers for the gearbox and differential were mounted at the rear, one on each side with large air scoops. The rear body work was fitted with a boot spoiler to induce aerodynamic down force and increase high-speed cornering stability and grip. An external 'dry break' quick-fill fuelling system allowed quick refuelling during the race.

The chassis floor pan was strengthened by a steel plate behind the lower sill panel with reverse-punched holes, allowing the plate to be welded to the inner sill skin and the central and front roll-cage footings. Broadspeed had not just fitted a safety cage but built it into the car for added stiffness to the chassis.

The glovebox was converted to give easy access to electrical wiring and fuse systems.

The engine rules allowed considerable freedoms, and a lot of time had been spent on the development of the base V12 engine including manufacturing new large valve heads with unique heart-shaped chambers, eliminating the original bowl-in-piston/flat head concept of the production V12. A Lucas mechanical fuel

BELOW
Left: Chassis strengthening.
Right: Glovebox of the XJC houses the relays and fuses.

Group 2 V12 engine producing 450-480bhp on mechanical fuel injection.

injection system was used with an injector fitted above each of the 12 individual intakes and these were then grouped into four separate plenum chambers, each housing three adjacent inlet trumpets. The system was controlled by a butterfly in each of the four housings.

Oil surge had become another problem under extended braking periods with large amounts of oil flowing to the front of the long engine and away from the oil sump which contained the oil pump pickup pipe. This oil would often not be returned to the sump and oil pickup pipe in time to retain oil pressure when the car was accelerated out of the corner, resulting in many brief moments of very low oil pressure to the bearings, leading eventually to repeat bearing failures.

Because of the problem with engine-oil surge when braking and cornering, in 1977 a dry-sump lubrication system had been introduced in one car late in the team's uncompleted season. It had replaced the wet-sump system which was mandated under Group 2 rules and was at the very heart of the problems that had caused Broadspeed to withdraw from racing before the season's end.

We concluded that the XJC V12 was the realisation of a sophisticated engineering concept that had been expected to produce the best performance possible right from its very first race. The Broadspeed team had not expected to climb a learning curve or to have to make improvements on the initial specifications.

They took reliability for granted in what was an innovative and complex machine.

Also, the idea of using the sophisticated XJC Jaguar to promote the Leyland brand we felt was a flawed marketing strategy from the start. The Leyland name was synonymous with trucks and buses rather than cars (and especially high-performance cars), so their marketing programme was based on a concept that had been poorly thought through; trying to promote the Leyland Group, as opposed to just Jaguar. To succeed, a Jaguar racing programme would have to work in the sole interests of the Jaguar marque, and not be diluted by any broader concept. The Leyland programme had been heralded with much hype and expectations had been high, and when it did not then deliver, the repercussions were hard for all involved. The hype had placed unfair pressure on the Broadspeed team with nowhere to hide when the results did not come.

Much later, in a conversation in 2010 with XJC driver Tim Schenken, he told me the car handled well but the brakes were a major problem. He could not continually race the car near its full speed, and if the brakes were to survive in a typical three-minute flat-out lap during a race he had to turn in a lap time of around four seconds slower to maintain acceptable brake temperatures. This negated the engine performance superiority of the XJC and the increase in lap times associated with the heavy race car and small brakes eventually produced unacceptable results. Two former drivers of the XJC also confirmed at this time to me the power outputs of 450-480bhp, the optimistic view at the time being 540bhp.

The question the TWR team now had to consider was how to overcome the obvious problems inherent with the XJS Jaguar, and what did John Egan need to hear to move this new programme forward? The specification and business proposal we then put to Jaguar was very different to that of the earlier Broadspeed XJC. From the very beginning variations allowed under the new Group A rules eliminated some of the problems. Transmission and brake freedoms provided for a choice of stronger gearbox and larger brake sizes. Just as importantly, a 300kg lower maximum weight, at 1400kg, would produce a faster, more reliable race car and reduce the load on the braking system. Although tyre sizes

The XJC driving compartment.

would be smaller than were allowed under Group 2, they would be adequate given the lighter weight allowed under Group A – although this would be a major task to achieve when starting with the XJS's production weight of 1750kg.

The two big problems remaining were the wet-sump lubrication system and the requirement for the car to look like a standard model straight out of the showroom. The latter prohibited the additional aerodynamic devices that had been available to the Broadspeed XJCs, and little could be done about this. We considered that either of these two problems could make or break the deal, so for my part, I paid very close attention to the Broadspeed engine's lubrication system and sump baffling, to compare it with the work we had already been doing through the TWR UK Rover programme. Meanwhile, the race team produced a basic specification to race an XJS, without the benefit of aerodynamic aids.

One year earlier TW had reassured Leyland's Austin Rover management he could bring them success, and we were granted the factory Rover effort for 1981. EngDiv. then had to rectify engine problems inherited with the Rover programme, with its previous team suffering from major failures when the BL Motorsport Rovers had begun their first year of racing in 1980. These engine failures were based around lubrication and bearing failures, and we had overcome these problems as the year progressed.

That new V8 engine programme had instantly more than doubled EngDiv.'s workload alongside the Mazda RX-7 programmes, still located in the same small shop in a corner of the TWR building. It also still had the same staff of two: Barry (Bazza) Holland and me, though we would be joined in mid-season 1982 by 16-year-old apprentice Charlie Bamber. Not surprisingly, that workload had resulted in countless all-night engine builds to cope with the many Rover engine issues in 1981.

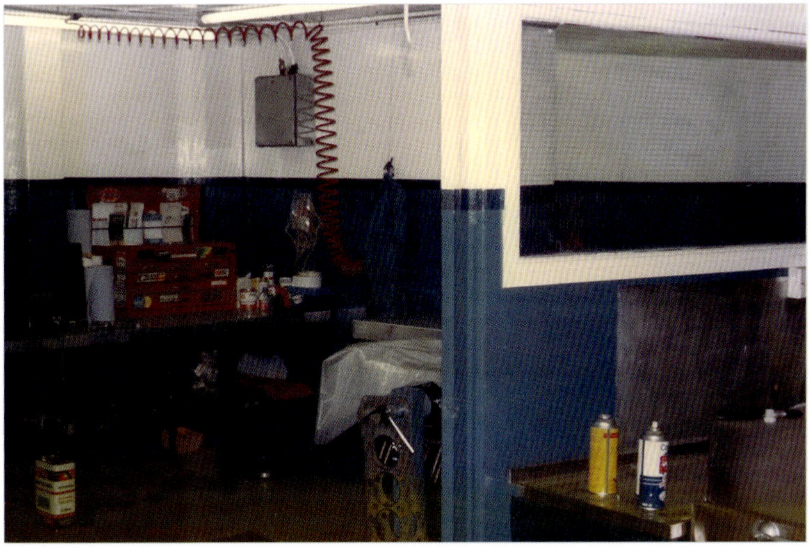

The engine division from 1979 till 1983.

TW now saw EngDiv., though it was still just a small part of his operation, as a possible trouble-shooter for manufacturers seeking to improve their competition prospects using their production based engines for racing. A major problem for us at EngDiv. was that TWR did not have an in-house engine dynamometer; engine testing had meant countless time-consuming trips to use the facilities at the BL Motorsport Competitions Department in Cowley, or Dr Erlich's facility at Milton Keynes where the Mazda RX-7 rotary engines had been tested and developed by us during 1980 and 1981 for the British Saloon Car Championship.

The Rover V8 engine problems were now mostly overcome by late 1981, and they had gone on to win their class in the BSCC that year, defeating the previously all-conquering Ford Capris. In TW's mind it was now time to move on to the next phase of developing his business, a racing project with a high profile name attached, Jaguar. We were all a small dedicated team struggling to keep up with his tireless energy and commitment.

1981 was an end to a year that had seen great expansion in the Oxfordshire-based company that had been running a single RX-7 race car when I had first joined in late 1979. Within the next two years it had expanded to four teams running a total of eight cars. These included the two separate Mazda teams, the two-car Audi team of Martin Brundle and Stirling Moss as well as the two-car factory assisted Rover SD1 V8 3500 programme. Thus we had been running three competing teams in the same series – the BSCC. It had been a hard but rewarding year, winning either class or outright championships in three of the four programmes the TWR teams had entered.

It was these successes that had moved TW from the level of an ambitious newcomer to see his company suddenly regarded as the leading team in British saloon car racing. To participate in the European racing scene was his next logical step, and he was ready, – even if his team was not completely convinced.

The XJS Jaguar looked a difficult but potentially very rewarding next step if the factory could be persuaded to back a programme in Europe. Fixing the lubrication system on the V12 would be a major concern given the previous history, but I would just have to find solutions, and with the little time available, the answers would have to be found while actually racing. There was little lead time and no money for a build-up year. At a meeting in late 1981 Jaguar chairman John Egan repeated Jaguar had no money for the 1982 programme, but we could have as many new Jaguar cars as we wanted at a 5% extra discount below the best price given to dealers! Something he no doubt was reminded about a few years later when TWR opened its first Jaguar Dealership in Coventry.

Egan and his fellow directors were very busy addressing the many problems with the revival of the Jaguar marque and their soon to be released XJ40 saloon car with its new 4 valve, 3.6 litre 6-cylinder engine, and the possibility of returning to motor racing seemed a difficult decision and therefore not a high priority. For the time being, TW seemed resigned to delaying this important next step in his plans, perhaps a mountain too hard to climb in 1982. At the end of 1981 he had said to me: 'Go off and have a good holiday: it's too late to do anything about Jaguar for next year.'

From my point of view as head of EngDiv. for TWR, 1982 would be a much-needed year of consolidation, enabling us to build up additional infrastructure and hire and train new personnel. The Rover programme was just under control, the Mazda was still very competitive, and TW had just assured me that a Group A XJS for the European Touring Car Championship series was off the programme for now, with 1983 looking more likely.

And so a few days later my seven-months-pregnant wife Karen and I relaxed on the short flight to Northern Africa en route to a very welcome ten days of relaxation in Morocco to escape some of the British winter.

TWR's investigation of the XJS

The first definitive design brief for the Jaguar XJS, or XJ27 as it was known was released as early as September 1968 by Jaguar aerodynamicist Malcolm Sayer. The major challenges were upcoming US regulations for emissions and crash resistance, which would mean the end for an earlier XJ21 concept which looked like an enlarged E-Type. The structure would have to be stronger and heavier, which forced the concept away from sporting to a quiet, smooth, fast continental tourer to rival Ferrari, Maserati and Aston Martin, albeit in much greater volumes and at Jaguar's usual low price.

Sayer had decided that this could be achieved using the same chassis floor pan as the XJ saloon, with a lower roofline and shortened by 171mm to 2591mm, using the saloon's suspension, engine and driveline, with the exception that a manual four speed gearbox developed from that of the E-Type V12 would be offered.

The only major difference in layout was swapping the XJ12 saloon's twin 11 gallon rear flank mounted fuel tanks for a central 20 gallon tank above the rear axle. The XJ27 concept styling was signed off by Sayer and Sir William Lyons in September 1970 and would remain essentially the same after Sayer's death later that year and Lyon's retirement in 1972. Doug Thorpe would finalise detail styling issues.

Prototype XJ27/1 was finished in 1972. The body shell turned out 14% lighter than the saloon at 327kg. Front end lift of 346lb at 130mph was found to be one of the first main problems in testing and led to the chin spoiler which reduced the lift to 225lb and also improving cooling and top speed by 2mph. Sayer's aerodynamics gave a drag coefficient of 0.39 compared to 0.48 for the saloon and 0.455 for the E-Type and this was all important information for TWR in a future race programme.

The design was signed off in 1973, and its development into a production car entrusted to Jim Randle, his first major project before taking over from Bob Knight as engineering director.

The Jaguar's suspension was an area that had produced the initial interest for Tom Walkinshaw.

The independent rear suspension [IRS] was standard Jaguar dating back in concept to the E-Type of 1961. It consisted of a steel sub frame cage carrying the diff and providing upper location for the twin coil/damper units per side. These units were mounted fore and aft of the drive shafts, which doubled up as the upper links taking power, braking and lateral and suspension loads through the inboard brake rotors to the differential mounted within the steel cage. The outer end of the drive shafts ran through bearings mounted in an aluminium upright casting.

The fore and aft location of these castings was supplied through heavily rubber bushed trailing radius rods going directly forward to the rear sill/foot well area. Lateral location and lower support for the four spring damper units was carried out by forged ends fabricated to a single large diameter tube between the hub casting and the suspension cage near the centre of the cars. The advantage of this system in production was that it could be fitted and removed as a whole. This entire unit was then completely isolated from the body by four metalastic 'vee-mount' bushes positioned to take the radial torque loads at each outboard end of the sub frame assembly.

The front cross member made up a complete sub frame unit with its attachments of suspension, brakes and steering rack as well. This used conventional forged upper and lower unequal length wishbones, with a single spring attached to the lower wishbone and its top mounted to the sub frame. The shock absorbers top mounted into the inner wheel arches. Once again this was all given limited degrees of movement via rubber bushing with the sub frame mounted on metalastic bushes to the body as well.

From an all out racing point of view however all this was less than ideal because it would mean the much higher suspension loadings produced by the racing tyres being used would not be controlled, with wheel geometry being inconsistent and compromised because of the uncontrolled movement of the suspension components.

Both suspension designs did, however, accept the large wheel widths allowed under Group A rules and still be contained within the car's outer profile. This was an important factor in TWR's evaluation of the suitability of the XJS.

The XJS introduced a new eight tooth pinion giving more direct steering, although it was still very light by normal standards and a decision would have to be made on its suitability in racing when using the power steering system.

In 1977, the auto changed from Borg Warner Model 12 to a tighter shifting GHM400, and the four speed manual was dropped in 1978 after only 352 had been sold. This meant that Jaguar did not have a suitable gearbox to go racing with, but fortunately would not be a problem under the Group A rules as any gearbox could be nominated.

Jaguar had then carried out an investigation of the XJS to race in Group 5 during the Broadspeed XJC period. This had involved wind tunnel work under the direction of Malcolm Oliver. Frontal lift had been a major problem and Abbey Panels was commissioned to produce a lightweight version for further assessment. This was all terminated with the Broadspeed programme's demise.

Lucas/Bosch P fuel injection had been introduced in 1979, increasing power to 295bhp at 5250rpm and torque to 317lb/ft at 4000rpm. In 1981 an improved version of the XJS was powered by the May-head High Efficiency engine, which took power up to 299bhp at 5000rpm and torque to 318lb/ft at 3000rpm and was another important factor when considering the racing proposal, with two engine configurations to consider, but both having the benefit of fuel injection.

The car had however been launched into a fuel crisis, and its profligate fuel consumption and unresolved styling had led to disappointing sales, falling to 1057 cars in 1980. This had led new chairman John Egan to completely suspend XJS production, although sales were never officially halted.

This was the reason therefore that Jaguar was now considering TWR's proposal to increase XJS sales and Jaguar's image via a return to the race tracks of Europe.

chapter two

A new direction for TWR

After returning from our brief holiday in Morocco, we had planned to travel up to Aberdeen to spend Christmas and New Year with friends we had met on our arrival in the UK in 1979. However, before the long drive to Scotland I dropped into TWR to reassure myself that all was shipshape at the factory. I needed to be sure everything was organised to start the new year. To my amazement, on arrival I was greeted by the sight of race mechanic Kevin Lee weighing a Jaguar XJS steering wheel horn centre. I then noticed a partially dismantled white Jaguar XJS sitting in a corner of the crowded TWR workshop. What was going on? TW had said Jaguar was out of the picture for 1982.

'Two pounds,' Kevin said, smiling referring to the steering wheel centre, 'and the heater is over 50 pounds; the weight reduction programme might end up easier than we thought.'

Originally built to Japanese specifications, the white XJS had been unsalable owing to some manufacturing fault. For more than a year it had been lying around the Jaguar factory at Browns Lane, where it was used by the Experimental Department for tests. Now, after £6000 had changed hands, this reject XJS had found a home at Kidlington, where it was set to become TWR's challenge vehicle for the 1982 European Touring Car Championship [ETCC]. This all came as a bit of a shock to me but certainly should not have been a great surprise. I had learned that after just over two years with TWR, anything was possible with our boss.

A quick run upstairs for a meeting with him revealed that the French oil company, Motul, had come up with some product and money. TWR was already Motul's official British distributor, through our sales department which was run by Hugo Tippet. This relationship had its beginnings in the Belgian National series with the TWR Mazda RX-7s in 1980 and Motul's introduction to the team via Bernard de Dryver, the number two in the TWR Belgian team.

TW had taken over the deal with Belgium importer Leo Van Haecht by the end of that year much to de Dryver's disappointment, and he had countered TW's actions by pulling into the pits during the championship deciding last race with a non-existent problem, before rejoining the race ensuring the Championship was lost. Motul had now been persuaded to venture into a European series with TWR. Significantly, Motul was the world's first 100% synthetic oil, its product 300V, gave a significant advantage when racing with wet-sump lubrication systems from its superior lubricating abilities.

First XJS – 1982.

The Jaguar factory had agreed in principle to instigate a return to motor racing, initially at arm's length, and the XJS race car was to be an unofficial Jaguar backed entry. TWR would be financially rewarded by Jaguar on a results-only basis. Jaguar would pay £30,000, £20,000 and £10,000 for first, second or third place respectively – but TWR had to finance and run the programme. Jaguar were counting on publicity in the sporting press to promote the emergence of a revitalised Jaguar Cars; with a substantial part of the funding available from them, provided we won. TW had reasoned that if Jaguar would back him unofficially for now, they would come to the party officially in future seasons. It was a huge gamble that had received the final green light when he had also managed to strike a deal with the BSCC Capri and Mazda RX-7 driver Chuck Nicholson. Chuck was a member of the Nickerson Seeds family, a well known UK business enterprise, but was also independently wealthy, with many businesses in his own right. Tom had convinced Chuck he was better off spending the money he would normally allocate to his entry in the British series on sharing a drive with him in a Jaguar XJS in the ETCC. Privateer BSCC budgets were typically around £70,000 for a one-car team at that time.

Most people in the same situation would have trumpeted this new deal in the news media for all it was worth, but TW was acutely aware of the pressure that had been put on the Broadspeed team by too much advanced publicity. For now he would put aside the personal kudos of gaining the semi works deal, preferring a low-key start to the ETCC. There would be plenty of opportunity for publicity later if we won – but he would also have to shoulder the responsibility and save face for Jaguar if it all did not turn out as planned. The rules were simple; if the car won, much of the success would be down to Jaguar's engineering assistance; but if it lost, it would be TWR's responsibility alone.

Our earlier investigation of the disappointing Broadspeed XJC V12 programme showed the press build up soon had a negative effect on the Broadspeed team when it did not win straight away. The PR machine had convinced the public that the powerful XJCs would stick it to the Europeans, especially the BMWs, from the start.

We had earlier been told there were discussions with Broadspeed, or even directives from the British Leyland PR department, to reduce the size of the front brake ducts so the large LEYLAND decals stipulated in the contract could be displayed. This type of interference was something the TWR team would now be able to avoid in its first year with its considerable level of independence. TW had therefore decided that 1982 would be a TWR year, with Jaguar Cars keeping themselves at arm's length.

All this had been sprung on me rather suddenly and the 1982 year was now looking very different, but there was another surprise still to come. TW continued by telling me that the company was now finished with the rotary-engined Mazda that had been my original reason for joining TWR. His new focus on working with British manufacturers was influenced by our past experience with Mazda which had shown up communication difficulties between the two different cultures. TWR had found itself unable to control that programme while negotiating the factory co-operation needed when working with Mazda engine detail for Group A. It also did not help that Japanese vehicles were restricted to sales quotas in Britain and Europe with Mazda allowed a quota of about 10,000 a year. The intention here was to protect the European car industry,

Co-funder Charles Nickerson raced and is known under the *nom de plume* Chuck Nicholson to avoid the family connection.

and this restriction had led Mazda to be more focused on the open markets of the USA and Australia.

The XJS race programme satisfied the idea of TWR 'going British' in the future, working with a culture it could understand and effectively control. The change to the V12 engine programme would take up all the time released by losing Mazda – about half of my time. TW's engineering logic was now going to assign half a man-year to a new V12 engine project that was fraught with difficulty! TW had driven against the Broadspeed Jaguars in his days as a BMW works team driver, but he was now about to promote the Jaguar brand to the UK and specialist press around the world on Jaguar's behalf.

In 1982 TWR would field a single XJS in the ETCC, and continue with a two-car Rover team in the BTCC. Jaguar would provide EngDiv. with two of its second hand standard V12 test engines, making a total of just three for us to build and use for the entire season.

The European series would start in three months time and include the Spa 24 Hours endurance race just three months later – although at the time, the first TWR V12 engine design wasn't even started, let alone built, and the race car was just a semi stripped road car. And that was the sum of the Jaguar XJS deal that TW had calmly explained to me.

The fact that TW proceeded on this basis with Jaguar was an early indication of the determined and far-sighted way in which he wished to operate, as well as the gambler in him. Outwardly at least, he was not daunted by the risks, and there is no doubt he was a big risk taker; and impatient for personal success.

Kevin Lee working on the first body shell.

There simply was no other person in Britain at the time that would have done the Jaguar racing deal on the limited factory support and terms offered, let alone turn it into a successful programme. These early days would be crucial to both companies as the foundations of a strong relationship were becoming established. The eventual outcome though would exceed everyone's expectations; even TW's, I suspect.

His business ideas had developed and come to the conclusion and philosophy of building a company that would enable British manufacturers to remain competitive in European motor racing, fast becoming the domain of German manufacturers. He was also able to present a complete package to Jaguar that now included engine design and development – this second string to his bow was something most teams did not have in-house. TW's British Manufacturer developmental policy though was one that few would have put their hand up for when you looked at the options available then. Much of the British car industry was in disarray owing to earlier British Leyland mismanagement and entrenched union mentality, and like most car manufacturers, was under threat not only from that source but an emerging Japanese car industry.

Most in the motor racing fraternity, when they later heard of TWR's plans to enter the ETCC with a XJS Jaguar, would give TWR little or no chance of success.

Things now became more frantic at TWR, with the build of the first race car commencing early in 1982, and the first race at Monza in Italy on the 21st of March. This seemed an impossible task with all the systems involved and was a long way from even beginning to look like having elements of an acceptable racing programme.

The new Group A regulations

From the outside these new Group A race cars would look just like standard cars, but now we had regulations that would turn them into high performance racing cars within their original exteriors. Group A was to be endurance racing for production saloon cars to worldwide FIA regulations. Each normal race must be a minimum of 500km, or a maximum of three and half hours. There was also one six-hour and one 24 hour race, so total reliability was vital. The race cars had to be based on standard models of which at least 5000 were built in one year, and fit specified internal dimensions to be accepted as a saloon car.

Outward appearance was crucially important with the race cars to look as though they were straight out of the showroom which was obviously good for new vehicle sales as Monday morning race results would be accompanied by photographs of a race car that looked just like those available to the prospective buyer in the showroom.

the very high 12.5:1 compression ratio with increased cylinder pressures from a better breathing race engine, while having to use the regulation standard road 97/98 octane fuel.

Jaguar's HE engine was not producing very much more power despite its new head design, higher compression ratio and use of fuel injection. I could not see any power advantages in using it; probably the reverse.

My final decision to use the earlier flat head design was assisted by the fact that the combustion chamber shape would be machined into the piston crown, something that could easily be changed and produced by an outside supplier such as piston manufacturer Cosworth. Potentially though not using the HE V12 version could be a political problem for Jaguar, but Jaguar had left us to make our own engine decisions and it was down to TWR to produce a race winner. This independence was one of the lessons learned from analysis of the earlier Broadspeed programme and which Jaguar had taken on board – vital to us all at this early stage.

On 14 January 1982, my thirty-second birthday, the other half of EngDiv., Barry 'Bazza' Holland and I set off for Coventry in two separate vehicles to pick up the two V12 engines promised by Jaguar Cars as part of their sponsorship package. We found our way to the Browns Lane Engine Test Department, and yes, they had heard about two engines for us; they were being used for corrosion testing around the other side of the building. I was impressed by their attention to detail, even though this was an all-aluminium engine. My naive thoughts quickly changed though when I saw two abandoned units lying under wooden planks in the car park. They had been lying there as junk for a couple of years after being used in testing programmes!

Regardless, 600kg of V12 engines were loaded into our faithful little Datsun 1200 van. They waved us through the security gate without checking for official papers – there were none – and Barry drove very slowly back to Kidlington. On the way a brief glimpse in my rear mirror revealed the Datsun's front wheels lifting off the road as it struggled up a small incline. Fortunately, Barry was an occasional weekend quarter-mile drag biker, and he kept most of the meanderings under control.

Having studied other manufacturers' developments of the bowl-in-piston design, I knew that this concept could tolerate fuel up to two research octane numbers (RON) lower for a given compression ratio compared to other two-valve engines, before reaching its detonation point. Group A regulations stipulated competitors had to use standard fuel, special race fuels not allowed. The engine compression ratio in Group A was unrestricted and it was important to be able to manage the best detonation control while using high compression ratios; this would give rewards in both power outputs and fuel economy. In the limited time available the combustion processes could be refined using the bowl-in-piston concept, and be cost effective for TWR/EngDiv. operating on a limited budget.

I then drew up a basic design for the piston crown shape, with bowl offset towards the spark plug producing a large squish land on the opposite side, and asked Cosworth Engineering to produce them – at £1200 a set. These pistons, however, could not be delivered until later in the 1982 season. With no time available to manufacture many special parts, and the forged Cosworth pistons being caught in a three-month-long production queue, our first race engine had to be very basic.

With it, we had to address two immediate major engine problems: oil surge under high G forces, and the biggest problem at this time, the control of the standard engine's electronic fuel injection system. This original Lucas/Bosch fuel injection system had to be retained under the Group A rules, but it was a sealed unit for exhaust emissions purposes, and no software to reprogramme the microprocessor chip existed outside of the Lucas/Bosch suppliers. Jaguar themselves were not party to this new technology and could not produce the technical support I had expected.

Now that TWR had the single XJS Jaguar in its workshop at Kidlington to progress the initial designs for the race car, Rover associate John Davenport, head of BL Motorsport, helped TWR to leak news about a racing Jaguar to the motoring press soon after. It was in a bid to placate the inevitable rumours that the motor racing industry would soon get to hear about, and perhaps come

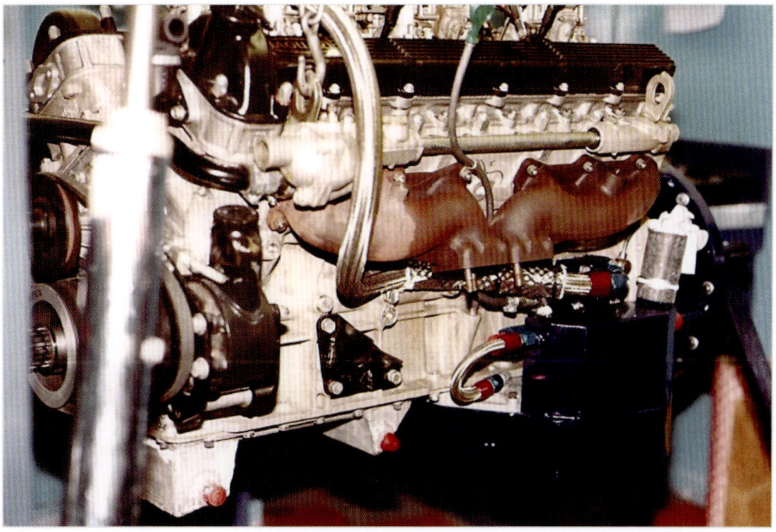
First Group A specification V12 engine.

to the correct conclusions. His communiqué stated that TWR was preparing a car for John Goss in Australia, primarily for the high-profile Bathurst 1000km endurance race and this story was casually allowed to gain credibility. Most if not all in the motor racing business thought that if TWR was going to Europe with a new Group A car, it would have to be with its Rover after the successes with it in the BTCC.

I soon realised after a couple of meetings in late January that assistance from Jaguar's engine division would be limited as it had not had an Experimental Racing Department for many years. They had however carried out a limited study and felt the standard engine was not far from the development potential under the Group A rules, and on open exhausts could well exceed 350bhp, but not much more. I was advised that the engine would not tolerate a compression ratio above 10.5:1 and that was about all the advice available at the time. While that statement would indeed prove to be true, the compression limitation would in fact be restricted by the production head gasket. To them, the magic 400bhp level that we felt was needed to win the series in the future looked out of reach. The need to eliminate engine failures was uppermost in Jaguar personnel's minds, and they put this ahead of encouraging higher power outputs at the cost of some inevitable initial engine reliability.

Jaguar had tended to trust what Group 44 in the USA had been doing with their racing V12 engines as they had been very successful while the Broadspeed produced units had not. However

Bob Tullius had become the 1977 USA Trans Am champion for Jaguar when Broadspeed had failed to deliver in Europe.

I visited Group 44's American base in early 1982 for an evaluation exercise on their Trans Am specification XJS.

There was little carry over to our Group A programme with the tube frame XJS weighing 1300kg and engine freedoms producing 500bhp.

this technology was of little use to us when building engines to the much narrower FIA Group A regulations, especially racing with camshaft profiles limited to standard valve lifts.

The lead time for the engine programme was extremely short, with the first race in March at Monza. Having no dynamometer facilities at TWR, I gratefully accepted Jaguar power train manager Trevor Crisp's suggestion to contact former Jaguar employee Ron Beaty, who owned and operated Forward Engineering at Barston in the British Midlands, and they had their testing facilities available to us.

There was no easy solution available for reprogramming Jaguar's Lucas/Bosch fuel injection system, so Ron Beaty then lent us a six-carburettor downdraft IDA Weber system, a set of re-ground camshafts, as well as the large capacity baffled oil sump from his offshore racing boat engine. These items were fitted to one of the 'pre-HE corrosion test' engines, while the heads, still with standard valves, were re-profiled to improve air flow and assembled using Beaty's Terry's valve springs and mild camshafts.

The rest of the engine was rebuilt using standard components, including cast pistons, complete with 1960s-technology large-section compression rings. These rings were subject to the onset of ring-flutter at 6500rpm, and would later lead to piston land damage. The V12 engine's ignition system was upgraded to the latest Lucas constant energy system used on the HE V12 version with a single coil delivering a high-tension spark to the distributor cap and a second coil wired into the system but with its high-tension outlet sealed. Timing was triggered by an electronic pickup and a six-segment star trigger wheel located in the distributor. The distributor itself was housed in the central V of the large V12 block, and came from the updated HE V12 derivative as well. It superseded the earlier Opus system, and would continue to be used throughout the entire Group A programme and it would prove to be able to keep up with future racing demands.

In the meantime the race team had been hard at work completing all its planned modification to the XJS in time for the first race. The basic Jaguar weighs about 1750-1800kg, which was

Baffled sump with side breather tank attached.

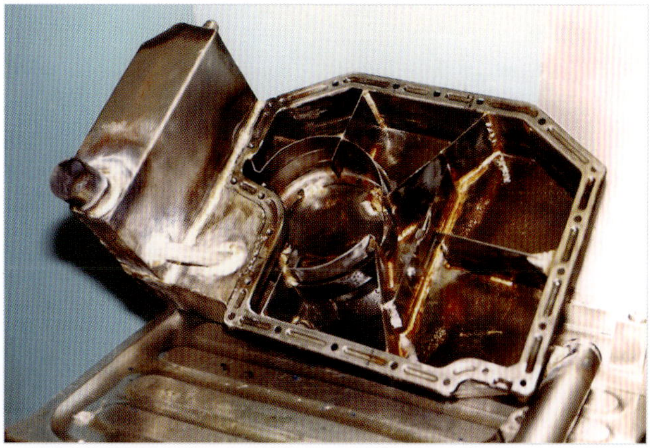

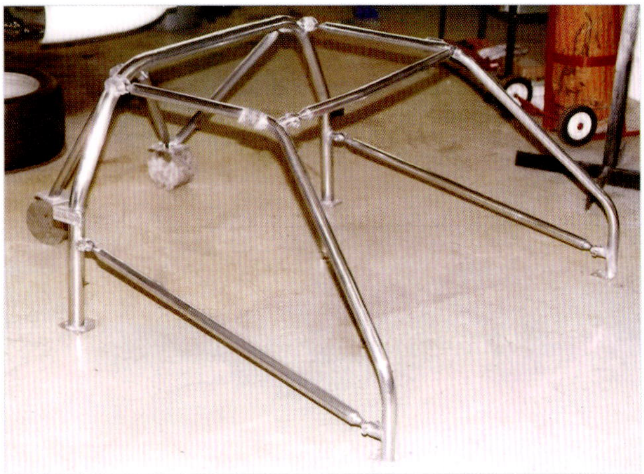

Aluminium roll cage.

350-400kg over the minimum Group A weight the XJS was allowed to race with. Body panels however could not be substituted for alternative materials; it was a matter of paying attention to detail with every item that could not be removed, and discarding all items not needed. The car's weight reduction programme also had to contend with the weight increases from larger brakes and wheels, plus safety features dictated by the regulations. A full safety roll cage made from aluminium tube was bolted to the interior. This cage had its interconnecting ends fitted with adjustable rose joints and allowed the system to be adjusted and preloaded to add an element of structural stiffness.

Attention to the factory spot welding of the whole car was carried out and additional welding was introduced in critical areas to increase chassis strength. A large 120 litre crash resistant fuel cell was fitted into the rear boot compartment area, as well as a 5kg BCF fire extinguisher plumbed to deliver its contents to vital parts of the car and driver, all adding additional weight. The interior was cleared of all inflammable materials and seating was replaced with one lightweight racing seat to contain the driver when subjected to high cornering forces. A special racing harness was anchored to the floor at six different points with over shoulder, side and crutch straps for restraint in the event of an accident. The mandatory battery isolation switch was fitted, with easy external access for rescue teams.

The XJS was completely rewired by John Smith from Lucas and this relationship would continue for the life of the entire XJS programme with his expertise adding reliability in a potentially difficult area.

Stripped interior showing new wiring carried out by John Smith of Lucas.

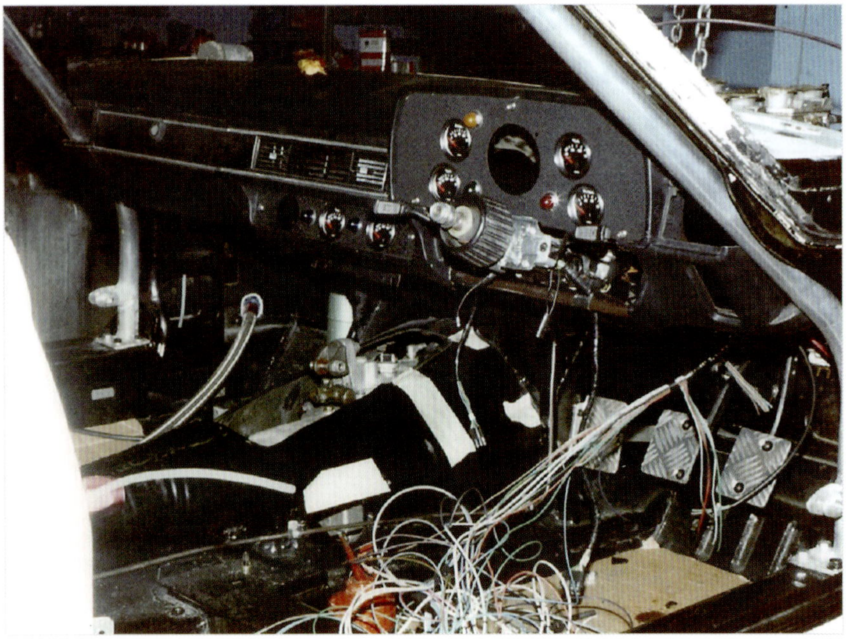

Homologation

Under the Group A rules factory specifications had to be adhered to except where the rules stated freedoms were allowed. That required measuring and noting of basic dimensions and designs on parts of the XJS that had to remain standard. Everything had to be photographed and the documents assembled and presented to the FIA in Paris before the recognition papers could be issued – and by now the first race was just four weeks away, in Italy!

This homologation process is usually carried out immediately by manufacturers who think a new model might compete in FIA racing events when the vehicle first goes into production. To date Jaguar had only progressed to a very basic standard because it had not even thought of racing the XJS when it was first produced in 1975. However, Jaguar had completed some very basic documentation work in 1979 for John McCormack in Australia to enable his GT category XJS to race in the popular national series there and this was a start.

Ron Elkins, who worked for BL Motorsport, had earlier co-operated with TWR on its Rover homologation programme and was familiar with these formalities. Fortunately his expertise could now be seconded to Jaguar to manage TWR's needs, since Jaguar and Rover were still part of British Leyland. Ron would eventually stay with Jaguar and later become their full-time liaison officer to the TWR Group C team.

One of Ron's tasks was to proceed with full XJS homologation for Group A racing, a process which produced a few more headaches compared with the simpler needs for standard production certification. Group A rules were designed for mass-produced saloon cars of four seats minimum, with designated internal dimensions and a minimum production figure of 5000 vehicles in 12 consecutive months. It was not a given that the XJS would qualify because the car was not selling that well and production numbers were not up to the needed level. Another problem was its description, that of a two-door coupe, not a saloon as envisaged by the rules.

The XJ6 and XJ12 saloon, not the XJS coupe, were Jaguar's main production saloon models, and the final acceptance process would turn out to be a close-run affair. Parts held in stock at Browns Lane were finally allowed to count as completed cars simply because 5000 units had not been built in one calendar year, just 3890 in 1982. Internal cabin dimensions, particularly the roof height above the rear seat area, were set by the rule makers to ensure 'saloon' cars only would be competing in this series, and the XJS needed all the help it could get in this area.

A half round metal bowl was used in the measuring process being placed on the floor with the rear seats removed and measurements taken from inside the bowl up to the roof lining. The car that was used for the FIA homologation process was to the 'Swiss' Jaguar specification, and this had the floor pan relieved in the rear seat area to allow the bowl's dimensions to sit lower. Fortunately the roof trim was also unique on this 'Swiss' car and that allowed the final measurements to pass the FIA test. This was just the start of a series of challenges to the rules that would become a feature of Group A racing for all manufacturers involved.

Further decisions on definitive chassis, transmission and braking systems had to be made, then photographed and submitted for FIA approval, if the XJS was to be ready for Monza. Once homologated, these items could only be changed on specific dates during the year and under strict conditions. Early in the year TWR's design engineer Eddie Hinckley was brought in from the Rover team to create the first suspension and front brake layouts, plus other items. These specifications had now been submitted and were awaiting the 1 April deadline from the FIA.

Strengthened front suspension.

Our evaluation of the Broadspeed XJC's rear braking system had shown a weakness in that particular area and I was well aware of the dangers when using Jaguar's inboard brake setup. The inboard system had caused problems on Jaguar saloons in long distance racing in New Zealand in the late 1960s. The amount of heat generated by the inboard rear brake discs, combined with their proximity to the differential, meant that heat soaked into the housing and raised the differential oil temperature as well as eventually damaging the output drive flange seals. The rear brakes were also located in such a way that they were difficult to service, meaning disc pad changes were impossible within an acceptable time frame when racing. For our type of racing these units needed to be fitted outboard within the rear wheels where it would also be far easier to duct cooling air to the discs.

TW, who was in charge of the development concepts of the race car, with their detailed application carried out by Paul Davis and his crew, took that idea on board and added this important feature to the list of modifications to be carried out. This would initially be introduced with caliper mounting lugs welded to the original aluminium uprights, and later a new rear upright casting was manufactured to add additional strength to the whole system.

This modification would also have another crucial advantage; it would mainly limit the stresses on the drive shafts to acceleration and compression forces by eliminating the fatigue inducing opposite-direction twisting motion on the drive shafts when braking with the standard inboard brake setup. This was a great improvement over the situation with the Broadspeed cars which had been troubled by brake overheating and drive line failures. Once again it was a question of pushing the freedoms to the limit, but with long distance races and fast brake pad changes in mind, moving the brakes outboard at the rear was an important modification. Interestingly this concept was later put into production at Jaguar in 1987 with its XJ40 saloon car. The upright assembly used in production is almost an exact copy of the TWR Group A casting.

Four Bilstein shock absorbers replaced the original rear units, each with individual coil springs. The existing Salisbury differential had many ratio options available and its limited-slip unit was retained as well. Strengthened lower rear suspension arms, and gusseted rear uprights were fabricated for the first race at Monza. The upper location for the rear suspension was formed by each driveshaft with a universal joint at each end, and their lengths would initially dictate the rear wheel camber settings. The lateral forces produced were fed into the differential and then to

BELOW
Left: Lower front spring mount.
Right: Adjustable front anti roll bar.

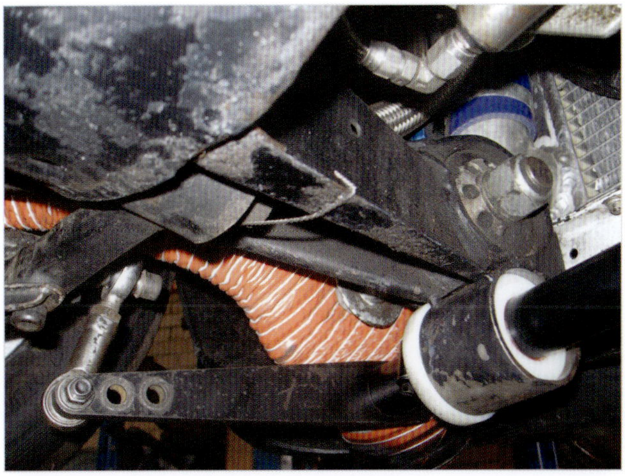

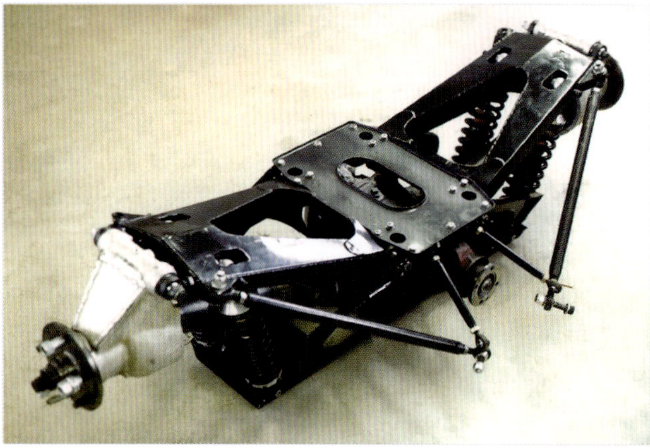

Rear suspension assembly showing fabricated lower suspension arms fitted later in 1982.

its cradle mounting systems, and this would need ongoing modification to maintain rear wheel geometry when cornering and accelerating because of flexing, as the first year progressed.

Strengthened and adjustable front suspension included replacement Bilstein shock absorbers, one each side and single smaller-diameter stiffer springs fitted in the original spring locality between the upper and lower suspension arms.

The suspension arms had fabricated mounting points added to take the shock absorber/spring assembly and move the lower mounting points outwards slightly to improve the original spring to wheel rate. The front suspension system produced a near 1.5:1 wheel rate where the suspension spring moved at a lower rate to that of the wheel travel, reducing some control and needing high spring rates to control the race car's suspension.

The original rubber mountings of the front cross member were replaced with solid aluminium bushes to remove flexing and help maintaining static suspension settings. The lower suspension arm inner mounting bushes were replaced with two adjustable spherical bearing joints and these were used to adjust the front wheel camber angles. A new larger front anti roll bar was fabricated with its arms attached to the chassis via two spherical jointed links. These arms had multiple mounting holes to allow variable adjustments to the roll stiffness.

Taped spherical bearings make the forward locating arms look like a one piece anti roll bar. Two mounting points on the under floor located the fore and aft movement of the rear cradle assembly.

Additional stay bars with spherical jointed ends had been mounted centrally on the chassis in front of the differential and attached to the lower central area of the differential cradle to eliminate fore and aft movement. The entire differential assembly on the original XJS cars was rubber mounted in a cradle that was then attached to the chassis via rubber mounts as well, and this had been discarded from the start as unusable. The rear suspension cradle was solidly mounted to the race car to provide a consistent platform to work with. Stay bars, one each side, attached to the ends of the lower suspension arms were

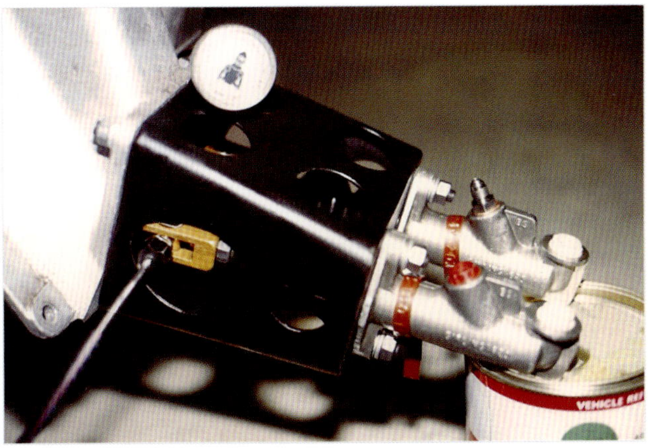
Driver adjustable brake pedal box used to change brake pressures to front or rear brakes.

fabricated and these angled forward and inwards to locate in the same central area as the cradle's central bracing arms. When these arms were joined via spherical bearings to their forward mounting points fixed to the chassis, and these bearings then wrapped in black tape, the system looked like an anti roll bar. This modification was then called a rear anti roll bar these being 'free' under the Group A rules.

AP Racing supplied 13in diameter ventilated front discs and four piston brake calipers, and these had individual hydraulic master cylinders for the front and rear brake circuits. Varying brake pressures front and rear were produced by a pivoting balance bar between the master cylinders and the brake pedal, and this system could be adjusted by the driver from a knob on the dashboard.

Special hubs for the soon-to-be-controversial single-nut quick-change wheel fixings were also designed, but machined by outside contractors as TWR had no in-house machinists or milling facilities. The need for quick wheel changes during refuelling stops led also to an onboard air jacking system being built into the four corners of the car. The individual jacks were activated by an air source plugged into a small hose coupling on the left rear pillar, and fed from air bottles stored in the pit area.

The single-nut hubs and wheels hadn't been homologated, or even considered in the rules, but with the obvious need for an extra stop for tyres and fuel with this heavier car compared to the BMW 528i, a Jaguar win would hinge on fast pit stops. Speed was essential, and the implementation of the single wheel nut and air jacking system was a typical response from TW in testing the rules to their limit in this first year with a new series. He was simply trying to redress any disadvantages the XJS might have.

Interpretation of the rules regarding single-nut wheel fixing would then be open to discussion from the first race at Monza, but by mid-year (and after some protests) a clarification ruling would go in TW's favour. All the other entrants then quickly adopted this system.

Only the four-speed manual E-Type gearbox, also fitted in limited numbers to the first XJSs, was available to TWR at this

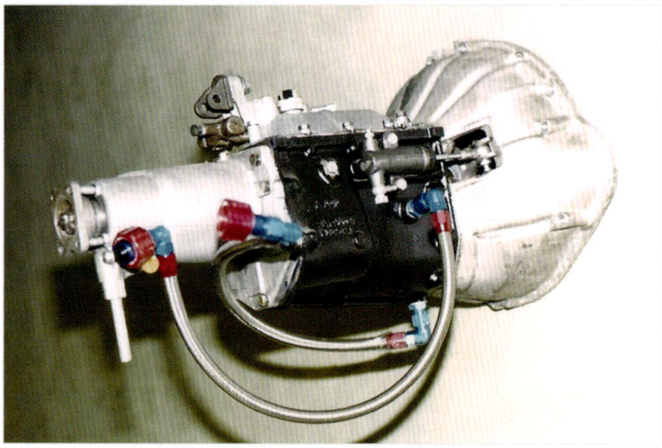

ABOVE
Left: Pump circulation system for the gearbox cooler.
Right: Close ratio 4 speed gear box with oil cooler hoses attached.

early stage, and Forward Engineering supplied one fitted with a close-ratio gear set for the first races. The gearbox internals were strengthened by fitting a double row of needle roller bearings to the cluster shaft. The gearbox options were free though and TWR had already planned to use a five-speed Getrag gearbox made in Germany when homologation was cleared later in the year. A standard Jaguar 10in clutch system was fitted and continued to be used in the early races before the 5 speed gearbox could be introduced. Both gearbox and differential were fitted with external pump-operated oil coolers, a necessity in long distance events. This first race car had the gearbox oil cooler cut into the front of the passenger's foot well.

The rules allowed large tyres; rims that would be up to 11.5 inches wide and 16 inches in diameter, fitted with tyres that were allowed to be up to 13-inch-wide. A sliding-scale formula for increasingly larger tyre sizes as engine capacity increased was another rule that compensated heavier race cars, but the maximum wheel and tyre allowed under this formula would not fit inside the original XJS rear wheel arches. This led to 'hand-finishing' of the outer wheel arch edges to make the body as wide as possible within the production tolerances nominated by Jaguar, but this still failed to allow the custom BBS wheels and tyres to fit within the rear wheel arches. Extreme hand-finishing of the inner arch would be necessary to avoid the danger of tyres coming in contact with body panels and puncturing.

Cutting 50mm out of this area and re-welding the rear inner wheel arch panel to accommodate the wheel assembly was covered by the fact that this modification had been carried out by Jaguar on two of its cars destined for the Middle East, when they had been asked to produce two special armour-plated vehicles. This modification became known as the 'Third world specification', and was controversial as it could be argued either way – and was, continually, by some in the early stages.

Modified rear wheel arch.

With the Broadspeed debacle in mind though, TW was not going to let any weak area diminish the competitiveness of his car. Without the option of increasing the clearance for the tyre by moving the inner wheel arches, a smaller tyre would have had to be fitted, not the larger one allowed under the rules, and smaller tyres would not last as long under racing conditions.

Other manufacturers also found ways to overcome similar problems. Measurements for body width Homologation were provided by the manufacturers, but also with their 'tolerance' allowed for mass production. If the FIA allowed a request for a one and a half percent manufacturing tolerance (which was a reasonably high tolerance for production purposes), this, when added to a vehicle width of say 1800mm, would allow an additional 27mm to be added to the overall width. Then the outer wheel arches could be stretched either side by 14mm (just over 0.5in), providing additional room for the larger tyres. The front double wishbone suspension on the XJS was however a positive aspect when fitting wider wheels when compared to the more conventional McPherson strut assemblies used by most manufacturers. These suspension struts were the limiting factor when offsetting larger wheels inboard to stay within the standard outer wheel arches.

Problems arose for all teams when the original drafting of the rules meant nominated wheel widths could not easily fit under the arches of most cars, especially as an increase in wheel diameter of two inches was also allowed. Nobody wanted tyre blow-outs caused by tyres rubbing on the inner wheel arch bodywork, and the controversial alteration in the rear of the XJS would be argued as a 'safety' issue, muddying the waters and adding some confusion to the legality debate. Safety issues must always be considered very carefully by controllers of motor sport. Most teams were massaging the regulations in this area but to a lesser extent. It would be TW's

The XJS with its first coat of black paint.

The new race car would start the season at an all-up weight of 1470kg without fuel – a fine effort by the race team. However, 56% of this weight was over the front wheels and this would be an ongoing problem with the XJS.

After the Goodwood test, the V12 engine was stripped and rebuilt, but the Cosworth forged pistons would still not be available for these first races. The Goodwood test engine specification became the basis for the early part of the testing programme, and a second engine was started in construction to be used in dynamometer development and become the race engine for Monza, fitted with electronic fuel injection.

Another test was carried out at Silverstone on March 2 with the car having been painted in black waiting final confirmation of Motul's backing.

One last outing in the multi carburettor specification came with the journey to Zolder in Belgium and a days testing with the German Bilstein company on March 4. This test was solely for development of their gas filled shock absorbers and various settings were tested and changed at the circuit using Bilstein's mobile workshop.

The six Weber carburettors had to now be replaced with the standard fuel-injection system and inlet manifolds for the first race at Monza, and the second engine was completed and sent to Jaguar at Browns Lane to be tested on their dynamometer.

Making the inlet manifold pressure based fuel injection system work with the new power characteristics gained from the open exhausts and changed camshafts was a major problem in the short time available before the debut of the XJS. We could adjust fuel pressure, but had no technology to remap the Bosch/Lucas system available to us. A high fuel pressure system was plumbed into the race car to produce 3 bar [45psi] fuel pressure to the electronic fuel injection system. This included low pressure lift pumps from the fuel tank to an anti surge header tank in the right hand rear section of the front fender (wing).

An anti surge fuel header tank was fitted to the recess at the rear of the RH front fender.

The Group A rules that mandated the use of production electronic fuel injection and inlet and exhaust systems, free valve timing with standard production valve lift, but gave freedom of compression ratios, were totally new to racing. No after-market interfacing systems to reprogramme production electronic fuel injection systems existed in 1982. This new electronic technology was the domain of a few large companies, and was focused on the need to reduce engine emissions to pass increasingly stringent standards in the USA for road cars; not motorsport.

With that in mind, I arranged a meeting with the Lucas Competition Department and they indicated the limited options available from them. The name of a former Lucas employee was casually dropped in conversation by the Lucas people, and it then appeared that some expertise might exist for a basic programmable prototype electronic analogue fuel injection system. Further discussions during the meeting about this person's input were quickly dismissed.

The Lucas Competitions option available was basically an electronic throttle cam based on the mechanical cam adjustment principle used in the control system of their mechanical fuel injection. This system would have a linear potentiometer running in contact with the throttle cam profile, and produced a varying voltage signal for the electrical control system as the throttle was opened and closed by pushing the potentiometer arm in and out. This changing voltage signal controlled the amounts of fuel from the electronic injectors by changing the injector open duration times. However, this system could not alter the important changing fuel demand through the wide open throttle rpm range. It was simply an electronic version of their two-dimensional mechanical fuel injection system. I really needed to find out more about that person who had hastily been dismissed in conversation. I sensed 'not-invented-here' politics were clouding the advice I was being given.

TW was pushing for mechanical fuel injection for the V12 engine based on his previous experience with the system on his BMW 530i, a project before my time at TWR. However, after these discussions with Lucas I was sure pure electronic control was the coming thing, even if it was not yet available. Besides, it was still far from certain that mechanical injection would be allowed in place of the standard electronic system as far as I was concerned.

I could not visualise such a fundamental change being permitted under Group A rules which stated original system to be retained, but could be modified. TW was always prepared to push rules to the limit though, and no doubt would somehow have forced through the acceptance of a mechanical injection system if we had become committed to it.

Avoiding this path would prove a decisive move for EngDiv. and indeed for TWR as we progressed with Jaguar's later racing programmes. The decision to pursue the largely untrodden path of electronic fuel injection would eventually lead to the introduction of total electronic engine management systems, and that was to prove crucial three years later with the move to Group C where engine rules would be strongly based on fuel efficiency, having to race with an allocated amount of fuel.

The first Group A TWR V12 electronic injected engine built for the season opening race at Monza in Italy was dynamometer-tested at the old Experimental Department at Browns Lane. We had the use of the test cell, complete with their Lucas electronic fuel calibration unit over the weekend, two weeks before the Monza race. The fuel calibration unit was simply a standard car control box with an external twist dial and numbers from 1 to 1000 to increase/decrease the injection pulse time, producing a corresponding fuel mixture change. It could be used to test and manually produce the correct amount of fuel needed at any given point, but there was no possibility of reprogramming and altering the varying shape of the fuel curve, something desperately needed with the changed fueling characteristics of the first race engine. Still, it was a start.

Various V12 enhancement programmes had been carried out at the Experimental Department by Jaguar engineers on their production V12 engine, and the most power seen in these experiments had been 350bhp, close to the fuel capacity limit using the production V12's standard injectors. Knowing these injectors had little in reserve, we fitted our first engine with larger flow capacity injectors from the 4.2 litre XK 6-cylinder engine. Various power estimates were put forward by those present [around 350bhp] before the V12 was started, trimmed via the unique fuel injection computer, and left for 30 minutes to run in.

The initial power test produced a barely driveable engine through the lower rpm range because of poor optimisation of the standard programmed microprocessor chip and the larger capacity injectors. Suddenly the power came on at 4000rpm, and the indicator needle promptly went off the end of the load scale. There were a few raised eyebrows from Jaguar personnel present, and once we had added more weight to the dynamometer control arm to overcome this calibration problem, it soon became obvious our engine had more power than any V12 previously tested at Browns Lane.

Jaguar's test operator became uncomfortable with these very different driving characteristics compared to the standard engine he was used to, and offered me the job of carrying out his testing. I was more accustomed to working with racing engines and their different demands when dynamometer testing and over the next 30 minutes began to make some sense of the preliminary set ups.

Using the conservative ignition settings of a maximum of 28° and with the best attempts at optimising the fuel settings with that calibration box, this first engine produced 390bhp at 6250rpm with a rapid drop off in power after that. These figures were adequate for a start, and we produced an operating window of around 2500rpm. The problem still remained in how to make this all work with the standard fuel control unit and how we were to run the car at the first race – with the transporter leaving in 10 days time.

After making various promises, I left with Jaguar's only dynamometer calibration box tucked inside my shirt because

TWR001 ready to go to Monza.

no official paper work could be raised for it to leave through the security gate during the weekend.

It was the first and the last time a TWR racing V12 engine would ever be run at the Jaguar testing facility. Plans were being formulated to build a single test cell for EngDiv. in Kidlington, but for the next six months we would use the facilities in a converted farm building at Forward Engineering in Barston to continue our testing and development programme. The team then fitted this tested engine and Jaguar's Lucas calibration fuel injection unit to the XJS, to be used at Monza in Italy.

At the same time the TWR Jaguar was revealed to the media in mid-March and it was announced to a surprised audience that TWR would run the XJS in the European Touring Car Championship. Everything had all come together in only three months, and been produced by a very small but dedicated group at Kidlington. The team's sponsorship from Motul, Akai, the results-based funding from Jaguar, and free Dunlop tyres, made this possible. The vital added patronage of Chuck Nicholson, TW's co-driver, should not be forgotten either, without this the programme might have never started – not in 1982, at least.

Maintaining a sense of humour had become increasingly difficult at TWR in the run up to the first race at Monza in 1982. I was being pulled in two directions, supporting both the Rover V8s and the new V12 XJS. Each team manager expected more of my time for their particular project, and that created a bit of a balancing act. TW was a great motivator within his small team at Kidlington, which consisted of about 15 people directly involved in both racing programmes, and all quite young. The average ages of the entire TWR team was probably 20–26 years, with managers in their early thirties. TW was 35 years young and the oldest employee, Hugo Tippet, was 39.

Ready or not, the team had to depart from Kidlington and travel the long distance to Italy and compete for the first time at the historic Monza circuit, the home of Italian motor racing. The solutions to any problems during this first foray would have to be produced by the team from the back of the transporter, a long way from its home base, in a foreign country.

Official announcement

837649 WALKIN G

5TH MARCH 1982

WHAT FOLLOWS IS THE TEST OF A PRESS RELEASE TO BE ISSUED SHORTLY BY TOM WALKINSHAW RACING AND EMBARGOED UNTIL WEEDNESDAY 10TH MARCH.

IT IS FOR YOUR INFORMATION ONLY.

PHOTOGRAPHS OF THE CAR IN ACTION WILL VERY SHORTLY B E AVAILABLE EITHER THROUGH CLIVE RICHARDSON IN BICKENHILL OR FROM ME HERE AT COWLEY.

URGENT

FROM JOHN DAVENPORT, BL MOTORSPORT

NOT FOR PUBLICATION UNTIL WEDNESDAY 10TH MARCH.

SUPERCAT GOES RACING

TEAM MOTUL JAGUAR ENTERS THE EUROPEAN TOURING CAR CHAMPIONSHIP

A PRIVATELY ENTERED JAGUAR WILL BRING BACK ONE OF THE MOST FAMOUS NAMES IN ENDURANCE RACING TO THE EUROPEAN CIRCUITS IN 1982.

A JAGUAR XJ-S H.E. HAS BEEN PREPARED FOR THE 12 ROUND FIA EUROPEAN TOURING CAR CHAMPIONSHIP BY TOM WALKINSHAW RACING LTD, OF KIDLINGTON NEAR OXFORD. IT WILL BE SPONSORED BY MOTUL, THE FRENCH OIL COMPANY FOR WHICH TWR ARE BTRITISH IMPORTERS AND DISTRIBUTORS.

THE 400 BHP V12 XJ-S H.E. - THE ULTIMATE SUPERCAT - IS SCHEDULED TO MAKE ITS INITIAL DEBUT UNDER THE BANNER OF TEAM MOTUL JAGUAR IN THE
~~FIRST ROUND OF THE UK CHAMPIONSHIP AT DONINGTON~~
FIRST UK ROUND OF THE CHAMPIONSHIP AT DONINGTON PARK ON 2ND MAY. BUT IF THE CURRENT TEST PROGRAMME IS OCOMPLETED SUCCESSFULLY THERE IS A GOOD CHANCE THAT THE CAR WILL BE RUN IN THE FIRST TWO CHAMPIONSHIP ROUNDS IN ITALY AT MONZE ON MA21ST MARCH AND AT VALLELUNGA ON 4TH APR APRIL.

WITH THE EXCEPTION OF THE GRUELLING SPA-FRANCORCHAMPS 24 HORUR RACE IN BELGIUM, ALL ROUNDS OF THE CHAMPIONSHIP ARE OF A TOUGH, FOUR HOURS DURATION.

A SECNOND XJ-S WILL JOIN THE TEAM IN TIME FOR THI S YEARS SPA 24 HOUR RACE ON 21ST JULY. TOM WALKINSHAW, OUTRIGHT WINNER OF LAST YEAR'S SPA 24 HOURS, WILL BE THE TEAM'S NUMBER ONE DRIVER. OTHER TEAM DRIVERS WILL BE: PIERRE DIEUDONNE, JEFF ALLAM, PETER LOVETT AND CHUCK NICHOLSON.

SUPERCAT HAS BEEN BUILT TO THE NEW GROUP A FORMULA FOR FOUR-SEATER TOURING CARS.

THE EXCITING PROJECT SPRANG FROM AN XJ-S DEVELOPMENT PROGRAMME WALKINSHAW WAS WORKING ON FOR AUSTRALIAN RACING.

'WE REALISED THAT THE XJ-S HAD THE IDEAL SPECIFICATION FOR THE NEW GROUP A FORMULA', SAYS WALKINSHAW.

~~[illegible]~~
'WITH A CAPACITY CLASS MINIMUM WEIGHT OF 1400 KG AND APPROXIMATELY 400 BHP ON TAP IN ENDURANCE RACING TUNE WE FEEL THAT THE CAR HAS A POWER TO WEIGHT RATIO EQUAL TO ANY OF THE POTENTIAL COMPETITION IN THE CHAMPIONHSHIP.

'COUPLED TO THAT THE STANDARD XJ-S HAS THE ADVANTAGE OF DOUBLE WISHBONE SUSPENSION ALL ROUND. THAT GIVES TREMENDOUS SCOPE FOR SUSPENSION EDEVELOPMENT WITHIN THE REGULATIONS AND ALSO ALLOWS CLEARANCE FOR THE MAXIMUM ALLOWABLE RIM AND TYRE WIDTHS, WHICH MUST BE ACCOMMODATED WITHIN THE STANDARD BODYWORK.'

in hemispherical combustion chambers, and a capacity of five litres. The result was a complex, heavy and expensive engine.

The first version of the quad-cam V12 engine design needed the inlet tract to be of a semi-downdraft configuration to fit between the V12's 60° block angle, but results were disappointing. After many modifications it still fell short of the best 6-cylinder XK BMEP (Brake Mean Effective Pressure) figures by almost 7%, perhaps indicating an increased level of parasitic losses as well. The 5 litre quad-cam prototype V12 produced 502bhp at 8300rpm using Lucas mechanical fuel injection, tuned length individual inlet and exhaust systems, 10:1 compression ratio, 1.875 inch inlet and 1.375 inch exhaust valves using a 72.5° half-period camshaft with .450 inch valve lift. This was effectively a full race engine with the best technology based on the XK paradigm of hemispherical combustion chamber with twin overhead camshafts and opposed valves. The disappointing results forced a rethink of how to make the V12 engine less heavy, complex and expensive with higher efficiency for its production specification.

This thinking at the time relied on volumetric efficiency and the belief that a hemispherical combustion chamber was the ultimate shape for a best combustion process. It did not recognise engine design concepts creating squish turbulence for improved combustion which could not be successfully carried out with this design.

One of the later concepts to improve performance considered by Hassan and Mundy included Harry Weslake's flat head inline valve design. This was a complete departure from the work carried out previously, and from all accounts was not well received by Harry Mundy. The bore size was increased from 87 to 90 mm in order to fit two vertical valves side by side on the centreline of the cylinder bore. With this design concept came the need for a totally new cylinder head design, each operated by a single overhead camshaft, not two as in the XK engine. This produced a simpler and cheaper camshaft and distributor drive arrangement. It also weighed less (each head was 12kg lighter) and being narrower took up much less space in the engine bay. It was this concept that Jaguar decided to put into production.

Following concerns with Lucas's mechanical fuel injection, a prototype electronic fuel injection system was tested. The results were promising but eventually had to be discarded because the manufacturer, Brico, decided not to go into production with this new technology.

Lucas provided their new Opus electro-magnetically triggered transistorised ignition system which was more durable and accurate than the previous low-voltage contact breaker points commonly used. Transistorised ignition would also provide a capacity that could easily cope with 12 cylinders running at the engine's designed maximum speed of 6500rpm. The initial compression ratio was reduced from 10.6:1 to 9:1, and allowed the use of the commonly available premium leaded fuel rated at 97/98 Research Octane Number (RON).

With this flat face head design a simple concave crown was incorporated in the piston design, making it simple and cost effective to experiment with different compression ratios. The one drawback was the valve size restriction, with the inlet and exhaust valves having to fit side by side within the cylinder bore. The XK head design had up to 1.875 inch inlet and 1.45 inch exhaust valves in its opposed hemispherical chamber configuration; but the new concept was limited to 1.625 inch inlet and 1.375 inch exhaust valves side by side within its 90mm bore.

This new style of head design however immediately produced increases in power output up to 5500rpm, and more importantly, reduced fuel consumption by up to 11%, while providing

Walter Hassan on the left with Harry Mundy.

increased torque output in the critical operating range.

Jaguar then decided to stop the piston 0.150in down the cylinder bore and use a flat piston crown with small depression, making the cylinder displacement 5343cc from the 70mm crankshaft stroke. This eliminated any possibility that valve cut-outs in the pistons might be needed, a fear provoked by concerns over trapping small amounts of fuel and the possible effects on emissions. That solution probably worked to an acceptable level when tested on the single-cylinder test engine because the offset port angle introduced the inlet charge in a circular motion to the cylinder and producing sufficient charge swirl. Engine design normally has the piston close to the head face at Top Dead Centre, [TDC] producing a squish motion to help create turbulence in the fuel/air mixture for improved combustion, and this was a flaw in the eventual production V12 design.

The single camshaft per head prototype V12 engine was fitted with two 1.75 inch Stromberg carburettors on each bank of six cylinders, attached to a tuned length induction manifold to enhance mid-range torque. Allowing for power loss (e.g., through silencing), in a final specification production car, this configuration gave 272bhp at 5250rpm, with maximum torque of just over 300lb/ft between 3750rpm and 4000rpm.

An enclosed top-deck upper surface, common to most engines, was replaced with an open-top block for easy assembly of the cast iron slip-fit liners. This would have a detrimental effect on basic block torsional rigidity, but would not be a problem in Group A racing. When fitted, the heads and sump gave additional strength to the block. The advantages of the added strength from the production engine's lower skirt design outweighed any strength loss with the V12's open top deck design.

Torsional tests showed a stiffness of 50,770lb per degree of twist on the new V12 aluminium assembly with lower skirt, compared to 24,700lb in the prototype V12 with the block lower surface on the crankshaft centreline. Although the early 5 litre prototype block cannot be verified as having exactly the same properties as the later production block, the design still indicated a substantial improvement, and one that would prove vital when we later used it as a stressed member in TWR's Jaguar Group C and IMSA race programmes.

Extensive use of aluminium castings throughout the V12 engine reduced its weight to the point that it now weighed little more than the final production iron block 4.2 litre XK 6-cylinder engine. With all ancillaries (including the air-conditioning pump), the complete weight was close to 300kg (670lb) – very heavy by today's standards though.

A novel idea for oil cooling in this engine involved returning the oil from the oil pump relief valve to the sump via an intercooler using water from the lower radiator hose as the cooling medium. It worked well in normal applications but when the engine was run for long periods at high speeds, the increasing oil temperatures resulted in decreasing amounts of oil from the pressure relief valve returning via the intercooler. This meant that the ability to control the oil temperature diminished just when it was most needed. It did, however, assist cold short running in winter by helping to initially heat the cold oil.

I met up with Walter Hassan, long after his retirement, in 1987 at Haynes publishing premises where the book launch for Andrew Whyte's definitive book *Jaguar Sports Racing & Works Competition Cars from 1954* was first released. Harry Mundy was also there but was very reluctant to consider that anyone could improve on his V12 design. It had been decided that a get-together of the old and new generations involved with the V12 engine, orchestrated by Peter Dodd, would be an interesting encounter. When Peter had asked me what power outputs TWR/EngDiv. were getting from the 1987 Group C engine, my reply of 735bhp was followed by Harry getting to his feet and walking off saying 'bull shit'.

During the investigation of this piece on the origins of the V12 engine, former Jaguar test driver Norman Dewis, perhaps one of the few reliable sources remaining from that period, insisted in a conversation with Tim Nevinson that Claude Bailey's part in the development of Jaguar's V12 engine has been overlooked. He also said that Walter Hassan and Harry Mundy were well known to the press at the time, but that Bailey's Jaguar peers credit him with much of the effort to produce the V12 engine.

The V12 engine was constantly upgraded during the 1970s, and the HE combustion chamber head, along with Bosch/Lucas electronic fuel injection was produced to upgrade its fuel efficiency and emission capability for 1981. It is with the introduction of the HE version in a facelift XJS in 1981 that TWR became involved with proposals to take Jaguar back into motor racing using the second generation XJS. Would this new look, better efficiency and claims of better build quality, combined with TWR's racing programme be able to rejuvenate Jaguar's sales and image?

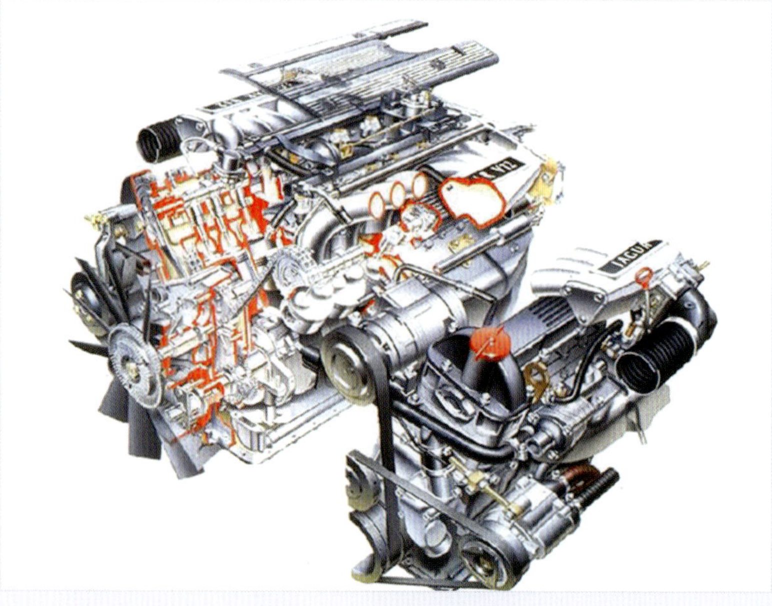

chapter three

The XJS takes to the circuits of Europe

In the first half of the 1982 ETCC season the single XJS Jaguar would compete in seven of the twelve races over a three-month period between March 21 and June 27, a very heavy schedule for a team with a new car. Each race would be a minimum of 500 kilometres long, with one race being held in Austria, Czechoslovakia, England, Sicily, and three in Italy during the March to June period. That would require TWR's large transporter with its full workshop, to depart on Thursday morning, and return to Kidlington by Monday night. The turnaround times at Kidlington between these races varied from three to nine continuous days – if there were no holdups getting back.

EngDiv. utilised Forward Engineering's dynamometer in the Midlands village of Barston for the running in and power testing of each engine. That meant a two-hour drive each way which was an added time burden, and these visits had to be fitted in with the engine removal and refitting schedule of the race team. The dynamometer was in an old farm building, with unsilenced exhausts pointed towards a convenient opening in the back wall. It was not the ideal situation but was our only option at the time.

XJS and transporter ready for Monza.

The TWR team arrived early at Monza and spent two days before the race testing systems and tyres. The driveability of the hastily constructed engine specification for the TWR XJS was not good, in fact it was bad. TW had to work to improve the car's handling and get the feel for tyres, compounds and suspension settings as best he could. Poor engine response improved with adjustments to the fuelling via Jaguar's loaned test-bed control box. The inlet manifold pressure based fuel injection signals were lower than normal with the wider period camshafts being used, and this was producing a rich signal as the throttle was opened. Better drive off the slow corners was gained by leaning the control-box setting, but this also meant running leaner on the straight at higher rpm than I felt safe with. Eventually we found a compromise setting which would suffice for the first meeting. It was a compromise we had to live with at this circuit, but was also the first indication that it was something the V12 engine could tolerate, an advantage when later being used in the fuel-efficient formula of Group C.

The Monza track has a fast entry onto its long main straight, and because of this and the ever increasing speeds of Formula 1 race cars it had become necessary to build the unpopular chicanes, tight snake-like twists fitted at the end of the straight, to slow cars. These new features had just been completed and immediately added further pressure to assumptions TWR had based its original concepts on. These new chicanes would push stress on brakes and engine oil surge to new limits when the cars were forced to brake to much slower speeds at this circuit than in the past.

Monza – Italy

Qualifying produced a morale-boosting second fastest time for Walkinshaw/Nicholson, 0.2 second behind the 528i BMW of Kelleners/Grano, the reigning ETCC champions. Interestingly a former TWR RX-7 Mazda was third fastest driven by the UK crew of Palmer and Blackburn.

In the Sunday race the Jaguar continued its impressive debut by building up an early lead which lasted to almost half-distance before problems caused by those new chicanes led to the XJS dislodging a gearbox cooler line, forcing a retirement.

The winners were the Swiss Eggenberger-prepared BMW 528is of Kelleners/Grano and Vanoli/Calderari, with the Juma-prepared 528i of Joosen/Vermeersch third. To give an idea of the

The Eggenberger prepared Kelleners/Grano BMW528i won the first Group A race at Monza.

cars competing in this new series at the first race, an Audi Coupe GT5 was fourth, and Alfa Romeo 2.5 GTV fifth and sixth, followed by the ex TWR Mazda RX-7, GTV-6, Chev Camaro, Opel Kadet GT, Audi GT5E and a Ford Capri. All except the Camaro came from the smaller class.

Later engine inspection with the sump removed showed some oil surge had occurred and part of the crankshaft windage tray, used to remove surface oil attached to the crankshaft, had fractured and eventually broken after contact with the crankshaft. This was reasonably easy to repair in time for the next race. Our hard work had made for quite a good start, but now everything would depend on just how quickly we could deal with the many small problems which needed attention and produce a car that could stay competitive until the end of a three-hour race.

Following Monza, my discussions with Jaguar's engine test personnel revealed that three different V12 fuel maps were available for their various markets. We now had an idea of the overall fuel curve needed, especially for increased fuel needs at higher rpm. After laying out the estimated shape, and comparing sections of the three different maps, Jaguar's Experimental Department engineers came up with a way to get closer to the fuelling needs for the modified engine. They reconfigured the only part of the software they had access to, the number of engine cylinders. The original programme with 12 cylinders was changed to 24. Nominating twice the number of actual cylinders fooled the control box as it counted the number of sparks the engine produced to ascertain the engine rpm, resulting in it thinking the engine was running at half the speed it actually was. The fuel injection signals coming from the software were shorter because of this. This additional leaning in the lower rpm range was compensated for by using the larger-capacity XJ6 injectors and the extra fuel needed as the engine increased in rpm could then be

The XJS at Vallelunga Italy.

to the left hand side of the sump and up into the breather header tank that was part of the sump pan, and then to spill out of the engine vent pipe. A considerable amount of oil was being held up in the header tank leading to oil surge at Vallelunga. Kevin Lee dubbed this and the next couple of races using the high oil levels as TWR's 'Torrey Canyon era' – a reference to the oil tanker that had been wrecked off Cornwall in 1967, leaving an enormous oil slick in the English Channel. The lesson learned here was that we needed a large additional tank to catch the oil from the sump breather and allow the engine to find its own operating level.

The XJS had improved in all areas, and at Vallelunga took pole position ahead of Kelleners/Grano and the Calderari/Vanoli BMWs. It finished its first three-hour Group A race in third place, with TW setting the fastest race lap. The winners were once again Kelleners/Grano with Eddy Joosen and Hans Heyer in second place. It was a great result at the end of just our second race; only sixteen weeks after the XJS had been purchased as a standard well-used road-going test car, and provided a much-needed boost for the entire team.

TW telephoned John Egan that night and told him we would have come second if not for what he reckoned was an oversight by the race organisers. An ebullient Egan immediately offered to pay as though we had officially been placed second, and so an invoice for much-needed funding to the tune of £20,000 plus VAT was sent to Coventry.

When the Vallelunga engine was stripped down it was found to have partially broken piston second ring lands, caused by using high rpm with the heavy-section standard ring pack on standard cast pistons. Fortunately, the breakages did not lead to internal damage, and was only found after both the Monza and Vallelunga races. Cosworth were still unable to supply the desperately needed

Hans Heyer/Eddy Joosen – two drivers who would later be part of the TWR team.

forged pistons and the lighter 1mm ring pack, and we were anxiously waiting for them.

Zolder – Belgium

Between Vallelunga and the next round at Donington, TW decided it would be a good idea to take the XJS to Belgium for that country's first Group A race, a non championship event at Zolder. As it turned out, he was correct, although the weary team needed some convincing at the time. The black Motul TWR XJS led from the start and took its historic first race win. While the field was not large at Zolder, our early win on this tight track was greeted by all with some surprise. Perhaps TWR was going to actually make this car work! Suddenly the side-line cynics became a little quieter and it was a much needed boost in the team's confidence.

Donington – England

After our problems with the cast pistons, Ron Beaty then offered me his forged Cosworth 10.5:1 compression pistons from his racing boat engine, and these were fitted for the UK round at Donington to the engine Kevin Lee later labelled 'Old Faithful'.

With a third placing in the previous Championship race at Vallelunga, the pressure was on at the first round of the ETCC to be staged in the UK and the home race for Jaguar.

With the UK BSCC series regulations changing to Group A in 1983, it was a time of observation, speculation and discussion for many of the domestic teams as they weighed up their options on vehicle choice for the following year. They also wondered just where TWR was going with the regulations, and whether the TWR XJS team would be competing in the UK BSCC series next year. Many even wondered if they should be looking at building a Jaguar too as the BSCC would not have an engine capacity limit under Group A rules.

If Jaguar was to come into the national series, TWR would have a year's development advantage, and with it looking close

AP Racing's lightweight 9.5in clutch.

much-needed improvement with the opposition already having them.

The gearbox change also gave EngDiv. an opportunity to come up with a new clutch to replace the heavy original Jaguar unit that had been used when the first XJSs built had been fitted with the four speed manual gearbox, the basis for the original homologation. AP Racing reduced the original ten and a half-inch unit to nine and a half-inch with a high-clamp-load clutch cover and four finger ceramic faced solid centre driven plate. This plate was fitted with a metric spine centre to work with the new gearbox. It improved reliability, and reduced the clutch weight by an impressive two kilograms as well as the lighter clutch plate reducing momentum drive and assisting faster changes with these synchromesh boxes. Group A rules stipulated we had to use the standard manual flywheel at its homologated factory weight, and this lighter clutch assembly was the only way to reduce the reciprocating weight in this area. AP's clutch worked perfectly throughout the entire XJS programme, and would later be used in the Rover programme too.

Brno – Czechoslovakia

Brno in Czechoslovakia was a spectacular circuit with a long racing history. It had continued to be open to the western world despite being in a semi-closed Soviet-controlled country, and had continued to hold its annual race for many years.

The event in 1982 had 50 entrants including six locals. Although it has now been replaced, the original public road track was a 6.8 mile circuit with fast flowing corners. It was a circuit that suited both TW and the Jaguar.

Additional fine-tuning to the engine had the ignition timing being advanced by two degrees and for the first time the XJS started to show some engine performance on this fast track. Qualifying saw the XJS beat the Kelleners/Grano BMW by more than five seconds, with the Czechs Vojtech/Enge in a BMW528 (having previously raced a Skoda in 1981) close behind. This was the first real wake-up call for BMW, and the TWR team was growing in confidence with each race start and the oil system now proving to be under control.

Walkinshaw leads away the start of the Brno race.

Right from the flag fall in the three-and-a-half-hour race, the Jaguar began to distance itself from the rest of the field. After TW pitted on schedule for tyres, fuel and driver change, Chuck Nicholson was able to rejoin still in the lead. That left the BMWs, which had started the race needing one less pit stop than the Jaguar, now on equal terms stop wise, but with no answer to the much faster lap times of the XJS.

Things did, however, begin to swing back in their favour when rain began to fall. Nicholson was told on the radio to stay out when it looked as if the rain shower would pass. Some of the BMWs had fitted intermediate tyres, as their first scheduled stop had been later than the XJS's stop, thus coinciding with the onset of the rain. This had enabled the Czech driven, Hartge prepared BMW to close on the leading TWR Jaguar which was still on slick tyres. Then, as the rain eased, Nicholson was able to pick up the pace and steadily pull away. Enge in the BMW then lost control and crashed into the barriers, his car out of the race. Now that we had a good lead and the worst of the pressure was off, TW decided we should play it safe rather than make the finish marginal on fuel. Nicholson pitted

A quick look during the last pit stop for the XJS.

'Nae grip Willy' was a constant theme in the early days from TW. Tom Walkinshaw demonstrating to Bill Mac what he would like from the Dunlop tyres.

Left to right: Allan Scott (author), Kevin Lee, Tom Walkinshaw, Paul Davis and Bill Mac.

close to the end for a quick eight-second top-up and continued on to win by 70 seconds. Delcourt/Swyssen came second, with Xhenceval/Micangi next and Kelleners/Grano gaining valuable points in fourth. At last Jaguar had their long expected win, but it was in a Communist country with no Jaguar dealers or market for their cars. They also seemed to think that winning was inevitable with the XJS but were very happy as well to send the next cheque.

Österreichring – Austria

The winning Jaguar was returned to Kidlington for its race preparation then transported to the next event at the Österreichring in Austria (also known as Zeltweg).

It was pole position once again for the XJS, but here the BMWs were much closer, and after one of the team's most challenging races the XJS had to settle for second place. The Jaguar had been very competitive in the wet conditions that prevailed during the race, but had to pit to fix a problem with the windscreen wipers, which cost it a complete lap and eventually the race win. Walkinshaw stayed in the seat throughout the race, and in a desperate come-back drive he finally snatched second place from Kelleners as they crossed the line just half a second apart and gaining an extra £10,000 in the process. The strong Vojtech/Gartner pairing were the clear winners by over one minute, with the Jaguar next, followed by the BMWs of Delcourt/Swyssen and Calderari/Vanoli in fourth and fifth places.

Nürburgring – Germany

It was another round trip to the UK and back to Germany in ten days for the Nürburgring event (at the original long mountain circuit), a very important race for the Jaguar factory and TWR. This time though, it was six hours of racing before the winner would be known, not three.

It was a crucial race in the British/German rivalry. Germany was a new target for Jaguar in its drive to increase sales, and logically, BMW did not want to lose face on their home ground with their existing or potential customers. It would be the Munich firm's most important race of the year. Broadspeed had challenged here and lost; for BMW the black TWR XJS V12 needed to be dealt with in a similar manner.

The Eggenberger team arrived with a third BMW for the very strong driver pairing of Hans Stuck and Dieter Quester, making a total of seven BMWs in the field of 45 starters in three separate engine capacity divisions. Compared with the four 528is which had started the first race of the season at Monza, it showed BMW's concern over the competitiveness of TWR's sole XJS Jaguar. It should also be remembered that in 1982 it was the first time the XJS had been to any of these European circuits, and there was very limited time to produce the best setup, unlike the local teams who were able to test at any time.

The daunting old 22.8km Nürburgring was a circuit TW knew well from his time spent here testing with BMW in the mid-1970s, but he was obviously on edge. The strain of financial commitments, unpaid bills, accounts were up to eight months behind payment, and the need to win on BMW's home ground to inspire Jaguar Cars into a future works' team contract, broke through TW's normal granite-like exterior. An open confrontation with team manager Paul Davis ended in TW apologising to him in front of the team. That was a smart move, and was appreciated by all.

At times like this the best place for TW was behind the wheel of a race car, and he set the fastest qualifying time – an incredible ten seconds ahead of the next 528i of Kelleners/Grano.

The lone British Jaguar established a comfortable lead on the first long lap, and later pitted on schedule to hand over to Nicholson, who accepted the challenge of his first race on this daunting circuit against drivers who knew it well. Impressively, he regained the lead from the three 528is that had yet to pit. The race looked under control until just after two-thirds distance, when Walkinshaw, now back in the car, had a puncture and returned to the pits at a slow pace from the 'Foxhole' 4.5kms into the 22km lap, to have a new front wheel fitted.

The XJS was soon underway again, but Hans Stuck put up a spirited drive to stay with the Jaguar, and positions changed a number of times until Walkinshaw finally regained the upper hand on lap 31. Towards the end of the race Stuck stopped for a short top-up of fuel, and after six hours, four minutes and 34.8 seconds Walkinshaw and Nicholson took a crucial win in this very important race. They had won by 65.5 seconds ahead of Stuck/Quester, with Kelleners/Grano third, and Herler/Doring (Opel Monza) fourth. Fifth place went to Vanoli/Calderari in the third of the Eggenberger cars to finish this race. With second, third and fifth placings for the Eggenberger team, this demonstrated the intense competition and Swiss reliability that TWR was up against.

After the longest six hours of any day that year, TWR had claimed a vital victory on German soil, and announced to BMW that Jaguar had arrived once again. Now the two-door coupe Jaguar was a genuine threat. However, when the team excitedly watched the race report on television in their hotel that night, local news showed only BMWs. There was not a single shot of the winning TWR Jaguar. The news simply stated at the last minute that a car from the UK had eventually won the race.

We continued to struggle with the ancillary systems on the V12 engine because of the under-bonnet heat effects on the standard Jaguar drive belt system, with temperatures reaching 110°C. Stop-start cycles resulted in heat-soak through pulleys, expanding them and thus altering the original belt tension by stretching. If the

Paul Davis had plenty to scratch his head about preparing for the Spa 24 Hours event. Note USA-style twin headlights.

fan belts loosened, any missed gear or over-revving would cause them to break or slip off. This continued to be a point of unreliability, with any driver over-rev causing belt failure. At this time I was building all the V12 engines as well as half of the Rover V8s for the two car BSCC team, and so stopping to redesign and manufacture a completely new drive belt system between races had to take a back seat, the drivers just had to be more careful.

Ongoing engine development continued as and when time allowed, and at this stage a specification for the Spa 24 Hours race had to be settled to ensure enough parts were available for a two car team. The demands had then doubled with a second race car now completed. This second car had been made from an acid dipped body shell in an experiment that Jaguar had carried out after the Broadspeed era – both Tom and Chuck would switch to this second car. The body shell had been sitting at Browns Lane for at least three years.

Working with Coopers Payen's Richard Weatheral, Jaguar's original equipment head gasket supplier, EngDiv. received an improved head gasket in time for its first Spa 24 Hours endurance race. The new design had special thicker gasket eyelet construction and new body materials which increased the clamp loading on the liners, and allowed the higher compression ratios to be used safely without leakage, improving reliability in this vital area. The increased eyelet thickness was also needed to compensate for the different expansion rate between the cast iron liner and the aluminium block that the heads were bolted to. With the aluminium's greater expansion rate the clamping load would decrease as the engine reached its operating temperature. After a couple of trials we came to the conclusion that an additional 0.004 inch needed to be added to a normal eyelet design to compensate for the entire thermal effect in this engine. Liner to block deck heights were kept at zero to 0.002in proud.

The second XJS for Spa takes shape in the TWR workshop.

Jaguar's power train director Trevor Crisp had told me their V12 engine would not tolerate a compression ratio above 10.5:1, which was true, but the limitation was actually Jaguar's head gasket. EngDiv.'s development of this new head gasket continued in Group A and would be of direct benefit to Jaguar some years later when its design features were used in the new 6 litre HE V12 production engine that TWR EngDiv. redesigned and developed in conjunction with JaguarSport. This new JaguarSport 6 litre unit would eventually become the standard production engine eleven years later – a direct benefit of race engineering.

We also manufactured heat shields to go between the inlet manifolds and the cam covers to reduce heat transfer from the very hot 110-120 degree oil circulating in that area. With the inlet manifolds absorbing this heat over time the inlet charge was also being heated and this reduced the volumetric efficiency of the engine when the air expanded, reducing the air density. The result was less air entering the engine under these extremes of under bonnet temperature.

For the Spa 24 Hours, and remaining races in 1982, we used a slightly narrower-period camshaft based on the L1 profile, while further fine-tuning improved the lower end of the power band, vital for running in the wet conditions expected at Spa. Two pairs of inlet manifolds had been cut open and carefully fettled by George Toth at Richard Longman's small engineering shop in Christchurch, Dorset. This helped improve the internal airflow and they were then welded back together again for this race.

The inlet manifold work had increased output to 400-405bhp at 6500rpm, despite the use of a 2° half period (8° duration) narrower camshaft period, and the power drop off at 7000rpm had been limited to 390bhp. While these were small power increases, on-track driveability was improving, with the fueling curve more suited to the improvements in volumetric efficiency at both ends of the power curve. These two areas had been slightly over fuelled and it was a matter of reversing the mechanical specification to better suit the fixed fuel curve we had to live with. The inlet manifold pressure based part throttle fueling settings could not be altered, but this new engine specification was working better with the standard production analogue fuel programme, but being read as a 24 cylinder engine.

Ignition timings were kept to a 30° maximum for this race, and these options together saw the first good step forward in engine power, economy and driveability. This work produced a more flexible engine and for the first time TW expressed his appreciation, feeling that the team now had a satisfactory engine performance level.

The two race cars took on a different look at Spa being fitted with the American specification twin headlights to help with the long night's drive.

Spa 24 Hours – Belgium

The TWR team arrived at Spa in Belgium with two cars for the first time. The driver teams were Walkinshaw/Nicholson/Percy in TWR002, with Dieudonne/Lovett/Allam in the second XJS, this car fitted with the old four-speed E-Type close ratio gearbox was the original TWR001 car, and 25kg heavier than the new TWR002.

The field had grown since the beginning of the season at Monza three months earlier, when just eight cars had been entered in our Division 3. At Spa no less than 10 BMW 528is had been entered, two 5.7 litre Chevrolet Camaros, three Opel Monzas and five Ford Capris – a total of 22 Division 3 cars in a field of 54.

The TWR team also bought along TW's old 3 litre Capri that had been lying in his barn, and after a hasty clean-up it was used for familiarisation laps by our drivers who were new to the circuit. The team could now work through set-up issues with both XJSs with drivers who knew the race cars, while the Capri would be used for some of the mandatory qualifying laps in both daylight and darkness for the others.

More importantly, we had been allocated three pit spaces in the crowded pit lane, having entered three cars, and while the Capri would start in the race, it was to be withdrawn at its first fuel stop.

Peter Lovett (rear) looks on in the crowded awning at Spa.

It would create an additional pit space should it be needed for working on one of the Jaguars during the race, as well as making plenty of room for fast entry and exit pit stops for the two XJSs.

The plan progressed well in qualifying until the night session, when Win Percy returned with no electrics. Inspection showed the wiring looms had been eaten through by mice in the barn, and hasty temporary repairs were needed. No instrument lights were available so a small torch was found and taped to the side of Win's helmet and he had then been sent out again. It was the only fix the team could produce in the limited time available, and was carried out in a good-humoured way. We were all glad to see Win back in the team and knew he could handle this situation with ease, although he did not seem as sure of it as we were.

The Jaguars qualified first and seventh, with all the spaces in between filled by BMW 528is. The five-speed XJS driven by Walkinshaw was almost two seconds ahead of the fastest BMW, and the original XJS with the four-speed gearbox, and its drivers racing the XJS for the first time, was within one second of the leading BMW group.

As is quite normal in this forest region, the race started in heavy rain, but Dunlop's wet tyres were, unfortunately, not working in these conditions. On the starting lap TW, in pole position, slowed to a halt on the start line in the deluge, and brought the rolling start to a halt by opening his driver's door. That amazing action showed his standing in Belgium; he was dictating to officials his belief via arm-waving to the starter that it was too wet to continue! The start was duly aborted. The real problem was that the XJS was undriveable on the Dunlop wets, although the weather quite suited the Michelin- and Pirelli-shod cars. Walkinshaw was desperately looking for any delay to gain extra time for the track to be drier.

The race was restarted soon after, with the rain continuing, and the Jaguars slowly slipped back out of contention and into the mid-field. It was left to the numerous BMW 528is to fight it out among themselves, and finally Walkinshaw decided to fit Pirelli wet tyres to both XJSs. That decision had taken some time, and was centred on the fact that the Dunlops were free issue, and a 24 hour race was not the best place to have to start paying for your tyres. The Pirelli wet weather tyres did transform the Jaguars and the lap times improved by around 15 seconds! This made life easier for all our drivers, but the press write-ups would not go down well back at Dunlop's Birmingham base.

Spa start line in the rain.

Wiring repairs and a hub replacement still kept both Jaguars well out of contention as the race continued into darkness.

Suddenly, after eight hours the TWR Jaguar team was down to one car when Pierre Dieudonne lost concentration on his entry lap to the pits while loosening belts and removing his radio plug. He took a short run onto the wet grass verge, lost all grip and collided with the Armco barrier. His XJS was out of the race.

Most of these wet tyres would all return to Fort Dunlop in Birmingham.

Heat inside the car during all races to date was a problem for the drivers and in this 24 hour race it was beginning to take its toll. Virtually no air flowed passed the rear of the engine and around the gearbox bell housing, and heat from the exhaust manifolds, which were close to the steel firewall and foot well, raised the cabin temperature above 50°C. The original fan-operated ventilation system had been removed to

A night pit stop with Walkinshaw closing the door.

save weight, which was normal practice in these racing cars, but it now became a debilitating problem for drivers who with double-stinting would be in the cars for up to three hours at a time. Exhaustion and dehydration was fast becoming a reality in the remaining XJS.

Soon after the first accident, the second Jaguar was involved in an accident in the dark foggy conditions, after a slower car turned into Nicholson's path and made heavy contact. Repairs were carried out, but with the suspension well out of alignment, the XJS was retired after putting in a few more laps to check its race worthiness.

The 24 hour race finished with the BMW 528is taking the first six places – Heyer/Joosen/Hahne, Jarier/Trintignat/Tassin, Kelleners/Grano/Muller, Guitteny/Bleynie, Cudini/Regout/Peltier and Delcourt/Swyssen/Baert.

Hugo Tippet.

TWR Jaguar's lap scorer and timekeeper, the quintessential British gentleman Hugo Tippet, later sent driver and car temperature information from the Spa race to experts at the US Air Force base in Upper Heyford for a report on the stress levels involved on the drivers during the race. They reported that all exceeded their astronaut endurance training regime, mainly because of the temperatures in the cabin. I distinctly remember touching the roll cage during the Spa race and it felt very hot.

The Spa result was far from what had been hoped for, and was not even a real test of the XJS's endurance ability as the cars had covered only a third of the distance. It would be down to the last remaining races for TWR to claw back some points, but the ETCC had effectively been won by the Kelleners/Grano team after their five wins in the first half of the series.

The last XJS is retired.

Silverstone – England

During the much-needed five-week summer break the Spa damaged XJSs were repaired and both race cars fitted with rebuilt 11.25:1 compression engines. It was decided that Pierre Dieudonne and TWR Rover drivers Peter Lovett and Jeff Allam would drive the second car for the rest of the ETCC series.

For the Silverstone race, team members enjoyed the relative luxury of being able to commute from home. After an intense season of travelling throughout Europe, this was the race where family and friends, Jaguar employees and enthusiasts alike, could attend to watch and cheer the re-emergence of Jaguar.

The Silverstone weekend had the added attraction of being the historic and prestigious Tourist Trophy (TT) meeting. Although only open to Group A racing cars, not yet the regulations for British based saloon cars until the following year, amazingly there were still more than 40 entrants who had made the journey from all over Europe. The Jaguars took the two fastest qualifying times, well ahead of the established European teams. There was however a new entrant here; a mechanical fuel injected BL Motorsport-backed V8 Rover in the hands of Australian expat Brian 'Yogi' Muir and sometime TWR driver Win Percy, who then qualified just a few tenths behind the second Jaguar. With TWR running

The Jaguars drew plenty of attention at Silverstone.

the works Rover programme for BL Motorsport in the BSCC, this was an interesting political development.

The XJSs would have to stop one more time than the BMWs for fuel and tyres as usual, but that would not be the only problem. Soon after the start the Muir/Percy Rover stuck to the rear of the Jaguars and all three cars pulled away from the rest of the field by two seconds a lap. Lovett pitted early for tyres and fuel on lap 30, staying in the car and rejoining just one place down, in third. TW pitted on lap 39, rejoining in second, and took the lead when the Rover came in for its first stop and lost time in its refuelling and tyre change to the well-drilled TWR crew.

After this delay to the Rover, and with Percy now in the driving seat, the BMWs were no match as Percy steadily regained the positions earlier lost.

With seven laps remaining, Percy was right behind Dieudonne in the second Jaguar when both suddenly slowed as they ran low

Bob Furn helps with the quick pit stop for the XJS.

ABOVE
Left: USA Presidents were soon to follow Tom Walkinshaw's lead.
Right: ... and what did you think of that John?

Left: ... I feel that for 1983 we have a very good package – TW the salesman.

Just the first one for your cupboard John.

on fuel. Dieudonne managed to reach the pits and rejoin after a few seconds of fuel was added, but Percy was out, his Rover unable to reach the pits, after an incredible performance from this team.

This Rover had proved it could be competitive in its own version of Group A specifications, and all options remained open for TWR and Rover.

And so for the first time TWR Jaguars took first and second places. It was all we could ask for in front of a home crowd – especially in front of Jaguar boss John Egan and the Jaguar factory personnel. And as icing on the cake, in third place the BMW of Kelleners/Grano was a whole lap down.

Zolder – Belgium

Two weeks later, the final round of the 1982 series was staged at Zolder, in Belgium. For this race the second XJS had been fitted with the five-speed Getrag gearbox, and Dieudonne had TWR Rover driver Jeff Allam as his partner. Allam qualified the XJS second fastest behind the Walkinshaw/Nicholson XJS on pole once again, and this was to set the scene for race day. The XJSs were never out of the top four places, depending on fuel stops over the first half of the race. After the additional fuel and tyre stop, the Jaguars re-established their leading positions during the last half of the race to have the Walkinshaw/Nicholson XJS win by nearly one minute from Dieudonne/Allam. They were followed home 40 seconds later by Joosen/Heyer and Kelleners/Grano in BMWs.

The results from the final two races earned us some very favourable publicity, and the press agreed that the Broadspeed XJC bogie had been well and truly laid to rest. They publicly recognised that Jaguar was a consistent winner, and BMW was in trouble.

The TWR Jaguar team finished its first year on a major high, with back-to-back wins in the last two races of the season, and two second placings. The two wins early in the season at the Nürburgring and Brno were as much down to TW's driving skill and sheer determination on very demanding tracks, as to car speed. Crucially, the last two wins showed BMW's 528i was no longer quick enough; BMW would need a new car for 1983.

BMW veterans Grano and Kelleners finished equal first in the Drivers' Championship, but Tom Walkinshaw was third. In the Manufacturers' Championship, Jaguar was second to BMW, initially using just the one car whereas BMW often had six. Surely this was enough to establish a TWR/Jaguar works team for 1983?

Season conclusion

The name that had been dropped at my meeting with Lucas Competitions earlier in 1982 was that of Bill Gibson, one of their former electronics engineers. His passion was electronic engine management, and he had been part of the Lucas team while the new HE V12 engine went through its fuel injection development and certification. After leaving Lucas, he had founded a consultancy company out of a tiny office in Birmingham. When I finally tracked him down he enthusiastically explained the programmable

The XJS showing tail up attitude under heavy breaking.

analogue fuel injection system he was developing, and said he could make it available to us. This all sounded very encouraging to me as it could bring us a huge benefit with the tailored control of fuel needs as we continued to develop the V12 engine.

Our modified standard XJS fuel injection system was controlled with signals from inlet manifold pressure and engine rpm, via a non-reprogrammable microchip which had been 'fooled' into a working position by telling it that the engine had 24 cylinders. The constant-energy ignition system and timing were both controlled conventionally by the distributor.

Gibson's new management system would initially control the fuel injection, but would use signals from the throttle opening angles and engine rpm, discarding the original manifold pressure sensing. The existing distributor-controlled ignition system would remain. Alternating between each bank of six cylinders, the six injectors would be fired simultaneously every 360 degrees; once with the inlet valve closed and once with the valve open, producing two injection periods over the 720° four-stroke cycles. Bill Gibson also felt that later it would be possible to have both fuel and ignition controlled from a full engine management system.

An eight-point (throttle angle) by 16-point (rpm) programming grid, giving 128 separate control points would introduce us to the new world of controlled electronic fuel injection systems. It was exactly what we were looking for all year to improve power, drivability and increasingly important, fuel consumption. For us at TWR's EngDiv. in 1982, this was the 'holy grail', even if it did have a few issues it could not handle in its initial form. Bill Gibson's analogue control board could be housed inside the standard Lucas/Bosch fuel injection control unit, which meant we would use the existing engine wiring loom and connections for water and air temperature correction signals.

It also covered the looming regulation clarification which would confirm that original engine management systems could be modified – but not replaced. This ruling came about after the entry of the mechanically fuel injected BL Motorsport backed Rover at the Silverstone race in 1982. BMW with their connections to Bosch, asked for the electronic injection ruling to be enforced, and so mechanical fuel injection was officially banned from the end of 1982. That ruling would give an advantage to European Bosch-supported teams who were confident that Lucas could not support the Jaguar in the same way because Bosch were licensing technology to Lucas, and could decide what form that could take. Lucas was not investigating its own electronic racing system in 1982, although that would change later in 1983. It was good news for us indeed that Bill Gibson and his one man business 'ERA' was now an option; and that retrospectively the decision to work with electronic fuel injection from the start on the Group A V12 engine became an even more vital one.

Late in 1982 I took a contract from TWR to Birmingham which acquired 50% of Bill Gibson's time for the benefit of TWR/EngDiv., at a cost of £6000, including exclusive support in all TWR areas of motor sport. That in itself was a welcome funding increase for Bill in the development of his system. Apart from some work with Brian Hart on his 4-cylinder turbo F1 engines, TWR was the only race team showing interest in Bill's business. To make ends meet, he had been forced to buy and sell used Jaguar parts while the electronic business grew.

By the end of the race season EngDiv. had, at last, completed the build of its own basic engine testing cell on-site, and by August no longer needed Forward Engineering's facilities, saving the four-hour round journey. Barry Holland updated the cell throughout the winter to enable us to continue the V12 engine development while new members were gradually being introduced to the EngDiv. work place – first of these being Peter Grace, formerly of Hesketh engines, in early 1983.

As the first Group A season came to a close, I began to use the prototype Gibson analogue fuel injection system on our own test bed, logging throttle and rpm points. We compiled the first detailed information for each grid point to construct a complete engine map for fuel using a Horiba CO2 meter. The testing information was relayed to Bill Gibson in Birmingham, who reprogrammed the analogue controls. While not perfect, with some flaws that had to be accepted because of the analogue programme, it was a huge advance for the engine programme at a cost of just £400 for each car unit.

By the end of the first season TWR had won four races with its XJS Jaguars, including significant ones in Germany and England. Surprisingly to most, it would soon be a step ahead of all other teams in terms of in house engine management systems thanks to the new ERA system. Virtually all forms of motor racing, including some in F1, were still using mechanical fuel injection plus a mechanical distributor or magneto ignition control. The potential for a complete engine management system, sometime in the future for us would be another large step forward.

TWR was still far from being well funded, and expenses such as developing engine management systems were not in the existing budget. It had been a very tight year financially with the results payment system from Jaguar. Additionally, TWR would be running Group A specification Rovers in the British Touring Car series in 1983, and the same engine technologies would be needed for the fuel injected Rover 3.5 litre V8. The BSCC Rovers had previously run on a single 51mm IDA Weber carburettor, but would have to revert to the production electronic system under the new regulations in the BSCC for 1983.

After discussions with John Davenport at BL Motorsport, he too soon understood the short comings in proposals put forward by BL Motorsport's engineers in the need to fudge and reprogramme the standard system, which had assumed Lucas production engineering help. John Davenport looked favourably on the concept of a totally new control system – but it would only be available through TWR.

In a meeting with TW and me at BL Motorsports (soon to be ARG Motorsport in its new headquarters in Cowley), Davenport offered a one-off £20,000 towards the final funding and manufacture of new programmable analogue units. This would be money well spent by Austin Rover and not a bad return on EngDiv.'s freshly signed £6000 contract, with individual unit supply to the Rover programme costing TWR £400 each.

Finally, towards year's end, word came through to us that TWR would become the works Jaguar team for 1983, although no official announcement would be made until the following year. This would be the first time since 1956 that Jaguar had entered a works team in international competition. The company itself was being re-energised, and now it was coming back to motor racing with TWR selected to provide all its race preparation and management.

For the first time however, Jaguar would not be using its own in-house resources to compete. Some of the older retired hands were not happy with this outcome, saying Jaguar should not compete if it could not do the programme internally, one in particular trying to stop it. By using the joint name TWR Jaguar, TW had added a second factory team to his rapidly expanding racing business after the TWR Rover factory team was established in 1981.

For John Egan and his fellow Jaguar directors, it was a declaration that the manufacturer was finally running its own business again, and had come out of the British Leyland closet whose umbrella business plan had almost destroyed it. Jaguar was raising its profile, and declaring its intentions as it had done in the early 1950s, through motor racing. Operating from the industrial estate at Kidlington near Oxford, the tiny but rapidly growing TWR racing business was playing a major role in assisting Jaguar's long-term future as a manufacturer.

The joint toe-in-the-water agreement between John Egan and Tom Walkinshaw in 1982 had been a decisive thing for both TWR and Jaguar in removing some of the pressure and producing a result that few had predicted. In 1983 things would be different. Additional pressure would come because press releases and statements would now be produced by Jaguar Cars after each test or race to assist and promote the Jaguar brand.

It all needed to be good news. Already, many at Jaguar Cars had pencilled in 1983 as the European Touring Car Championship winning year. It all needed to be good news.

At year's end Tom Walkinshaw was presented with the John Cobb Trophy, awarded for the best performance by a British driver, driving a British car.

This event was part of the prestigious British Racing Drivers' Club annual dinner and dance end of year awards held at the London Hilton on the Park. It was an encouraging end to a hard year for all at TWR.

Toasts

Her Majesty The Queen

☆

The Hon. GERALD LASCELLES
President, The British Racing Drivers' Club

☆

Presentation of the Club Awards and Championship Trophies

☆

Cabaret
Tom O'Connor
Rozanne Ward Dancers
11.30 pm

☆

Music
Johnny Howard and his Orchestra
10.30 p.m.–2.00 a.m.

The Club Awards

1982 B.R.D.C. Gold Star:	JOHN WATSON
1982 B.R.D.C. Silver Star:	DAVE SCOTT
The Richard Seaman Trophy:	JOHN WATSON
The John Cobb Trophy:	TOM WALKINSHAW
The Johnny Wakefield Trophy:	MICHELE ALBERTO
The Graham Hill Trophy:	JOHN WATSON
The E.R.A. Club Trophy:	JOHN WATSON
The Woolf Barnato Trophy:	SIMON PHILLIPS RAY MALLOCK MICHAEL SALMON
The Bruce McLaren Trophy:	VERN SCHUPPAN
The Nigel Moores Trophy:	MURRY TAYLOR
The Chris Bristow Trophies:	ALO LAWLER TOMMY BYRNE

End of year award ceremony.

CHAMPION

chapter four

TWR to be Jaguar's official works team

TWR entered 1983 and the second year of racing the Group A XJS coupe with the much-anticipated news that it had successfully encouraged Jaguar to make an official return to motor racing. Tom Walkinshaw would now take his two-car team into the ETCC with full factory backing and the status of a works team. The Jaguar deal would then complete his ambition to represent British manufacturers in motorsport, alongside the BL Motorsport (now ARG Motorsport) contract to run the works Rover Vitesse team developed to Group A specification for the 1983 BSCC series.

His business plan had taken just two years from concept to reality.

An additional building on the Kidlington estate across the road from the original TWR factory had been acquired, doubling the TWR team's facilities, and with the DART tyre business TWR now had three buildings at Kidlington. The Jaguar team's move to the new building had released pressure on the overcrowded original factory building that had housed both racing teams and TWR/EngDiv., allowing the engine division to expand from its small purpose-built area into that vacated by the Jaguar team. This resulted in a fivefold increase in working space overnight for EngDiv.

Long-term racing contracts are hard to come by, and expansion into larger premises based on a one- to three-year contract poses problems with their disposal if the race programmes are not renewed or replaced. The new TWR-Jaguar agreement saw the Coventry manufacturer take some responsibility for this potential problem. They would take ownership of any additional buildings needed for their race programmes, allowing TWR to expand but at the same time reduce some of the risks involved.

Jaguar had also produced a promotional booklet three pages of which are reproduced here.

Press release – excerpts from Jaguar 1983 European Touring Car Championship Programme.

Commenting on the return to Motor Racing Jaguar Chairman John Egan said:

"We watched and encouraged Tom Walkinshaw and his team last year when they scored four victories in their first season with the XJ-S. This year they will receive our formal support.

"Jaguar's image owes much to the racing programmes it undertook in the fifties and early sixties. In those days racing cars were directly related to the subsequent generation of road cars and this can be clearly seen in the evolution of the E-Type Jaguar from the Le Mans winning C and D types.

"The Group 'A' regulations for the 1983 ETC ensure that our participation will provide valuable spin off for the development of our XJ-S chassis and V12 engine over the coming years and indeed will give an added sparkle to all our other vital engineering programmes.

"We also believe that our ETC involvement will assist in our plans for sales growth in Continental Europe, particularly Germany which is the home of our major competitors. One of these, BMW, has dominated ETC racing in recent years and we know they have been working hard throughout the winter to prepare new cars for this season. Our two Jaguars will be trying hard to win!

"Whatever the outcome I am sure that this years ETC will provide enhanced spectator interest and will be a shot in the arm for European Motorsport."

TWR's founder and Chairman Tom Walkinshaw is backing Jaguar's European Campaign to the hilt:

"I've known how enthusiastic they all are at Jaguar for many years", he said. "John Egan, his engineering chief Jim Randle, and their fellow-directors have brought Jaguar to life again as far as I'm concerned.

David Boole, Director of Communications and Public Affairs continued on ...

For the purpose of Jaguar's racing programme a separate section has been set up at the TWR Kidlington, Oxford, headquarters. It will deal with the two racing Jaguars and nothing else; and it will work directly with Jaguar headquarters at nearby Coventry.

The cars will travel to the circuits in a distinctive white-and-green tandem-axle semi-trailer hauled by a fully-equipped Leyland tractor unit, and bearing the names of "Jaguar", "TWR" and "Motul".

The team will run two cars at most races, but not more. There will be six top-class drivers, and they will be selected by Tom Walkinshaw according to availability, race strategy, or any other factor that may obtain. Normally each car will be driven by two drivers during a race.

One car is being prepared from scratch, the other is (in updated form) the car which became the TWR 'No 2' car in 1982. Last year's 'No 1' car, which won four ETC races last year, will stay at base - available for development or, of course, for spares if these should be needed at short notice.

The Cars

The Jaguar is the world's only volume-production twelve-cylinder make of motor-car, and the XJ-S has established itself as a road car of the least-temperamental kind, with an almost uncanny combination of silence, power, and flexibility. Tuned for the optimum combination or performance and economy, the output has been boosted from the standard 299 bhp DIN to over 375 bhp. This has been achieved by the joint efforts of Jaguar Engineering and TWR's Jaguar engine specialist New Zealander, Alan Scott.

Alan Scott came to Britain via Australia, switching from race to rally cars. Then he met Tom Walkinshaw, and he is now in his fourth year with TWR. He has great admiration for the built-in strength of the V12, which (it is worth noting) went into production when the Jaguar director in charge of power units was one of racing's most famous technicians - the great Walter Hassan.

Group A permits the use of higher-compression pistons and altered cam profiles, but basically modifications are limited. For example, the regular XJ-S has electronic fuel injection, so the Group A car has it. Wet sump lubrication is, of course retained too, with suitable baffling to counter the exceptional forces which occur in racing conditions.

THE TWR SUPPORT TEAM

Paul Davis (TWR Jaguar Team Manager)has been with Tom Walkinshaw Racing Ltd for over 4 years ((since January 1979). He joined from the garage trade, having previously undertaken a civilian apprenticeship with the Royal and Mechanical Engineers.

His interest in motoring stemmed from his late father - a rallyist while serving with the RAF, and later an RAC scrutineer.

Paul has competed in autotests and rallies, and built his own specials for such competition work.

He is 28, married, with a young daughter.

Other TWR Personalities

Chief Mechanic	Kevin Lee	
Mechanic	Colin Marriott	
No 2 Car Mechanics	Philip Eberson	
	David Roberts	
Stores Mechanic	John Spears	
Engineering	Russel Sharp)	- Design
	Eddie Hinkley)	
	Alan Scott)	- Engineering
	Barry Holland)	

On 4 March 1983, one of the 1982 TWR XJSs, bedecked in a new 1983 livery for sponsors Jaguar Cars, Motul, Dunlop and TWR itself, was taken to the Waldorf Hotel in London for Tom Walkinshaw, flanked by Jaguar's John Egan, Jim Randle, Neil Johnson and David Boole, to officially announce to the motoring press that Jaguar Cars was returning to motor racing.

The sharing of engineering input that visibly increased with this agreement gave both sides increased credibility, and any small items that didn't suit a PR story were conveniently forgotten. One example was the much-hyped improvement in fuel efficiency with the recently introduced HE version of the production XJS V12 engine, and the need for the Group A XJS race car to have good fuel economy. This could have been a contentious issue because I had decided to use the earlier flat-head version of the V12 engine, with a modified bowl-in-piston arrangement, but HE badges attached to the race cars were all that were needed to cover that issue. TW had always insisted on bare car interiors, to minimise

Enzo Calderari, a BMW competitor in 1982 switched to the TWR Jaguar team for 1983, along with his personal mechanic Rolf.

The driver line up for 1983.

Tom Walkinshaw (G.B.) John Fitzpatrick (G.B.) Chuck Nicholson (G.B.) Enzo Calderari (C.H.)

Martin Brundle and Pierre Dieudonne with team manager Paul Davis.

EngDiv.'s new expanded facilities.

inflammable materials in the event of a fire, so no marketing person was going to insist on the trademark walnut veneer dash being retained, as had been suggested for the earlier Broadspeed XJC programme.

Jaguar management was happy to let TW run the show, and they would go with his recommendations; even driver selections would be his choice alone. This in turn would maintain good TWR-Jaguar team morale, being allowed to operate in a manner all were comfortable with. With these freedoms TW considered choosing a suitable multi-nationality driver line-up. Using some non-British drivers would ensure the publication of race results in their home countries and produce the much-needed 'free ink' that would promote the Jaguar marque. Additional means of funding were still needed and a few drivers would be happy to pay to join the larger factory-backed team.

There was, however, to be a reporting process and exchange of ideas and resources. Progress meetings were to be held at Browns Lane to ensure Jaguar was kept in the loop. The TWR team was for the first time subject to outside scrutiny, as well as suggestions for engineering support, and this was a new experience for our young management team. EngDiv. had limited drawing design capacity and no in-house machining facilities at Kidlington and being able to tap into Jaguar's Prototype Department for parts procurement and assistance would fill this gap. This support was to become increasingly helpful to us at EngDiv.

The 1982 all-black Motul colour scheme of the Group A XJS had gone, and two new livery schemes had been put forward for comment: a green base with a white flash, or white base with a green flash. It was then that another part of TW's character was revealed. Almost all voted for Jaguar Racing Green with a white flash, but TW's vote as over ruling chairman was for white with green. Many racing drivers are superstitious, and TW was not

Testing of the 1983 race car.

ready to test the bad luck that he felt might come with the largely green scheme.

The first XJS race car TWR001 from 1982 was cannibalised and a third new car built over the winter of 1982-83 using many of its mechanical parts. Perhaps understandably, the team was not too proud of the first car, which had been basically experimental, and they could not wait to remove it from the line up. It had copious holes and patches where things had been tried and changed, and was not to the build standard they now expected of themselves. Many improvements had been developed during the year, and with the 1982 experience behind them, they could now build a better car.

As the first race XJS was discarded and the team focused on improvement, no one had time to think that this race car might be a part of history, and worth preserving. Accordingly, after salvaging all the mechanical parts needed for the new replacement car, TWR001 was stored but then later cut up and scrapped. A lone voice at the time, the late Jaguar author and historian Andrew Whyte, suggested that some parts should be kept as mementos, or that some people might even pay good money for a piece of history. His pleas fell on deaf ears as we were all too busy, concentrating on performance-related improvements for 1983.

Some years later Andrew wrote to me and asked who was keeping the history of the team and its papers. I then became more careful with keeping records, and they have been invaluable in compiling my books on TWR's history after such a long period of time. Sadly Andrew died just three weeks before TWR and Jaguar were to win the Le Mans 24 Hour race in 1988.

EngDiv. at last had an effective development tool with its new dynamometer-testing facility attached to Unit 1 at Kidlington. A deal had been done with Brian Hart, a supplier of Formula 2 racing engines, to buy one of his surplus Henan Froude FO209 dynamometers with 90° cropped rotor. It cost £7000 with controls, and had been built into a room at the rear of the factory during the later part of 1982. Much of the construction work was done by engine builder Barry Holland, and its installation saved the long journey to Forward Engineering at Barston for testing our V12 engines.

Updates and development of the V12 engine over the winter break were made in parallel with the continued V8 engine development for the three-car Rover Vitesse Group A programme in the BSCC UK series in our expanded workshop.

The XJS had been very competitive in the final rounds of 1982, but the new BMW 635 CSi coupe which had been homologated for 1983 was a much more efficient aerodynamic design than the 528i saloon it replaced, having a lower drag and lift factors. Engine capacity was also up from 2800cc to 3500cc, producing over 25% more power. The Jaguar's aerodynamic principles had been developed ten years earlier than the 635i making it likely to be a much less efficient car on Europe's high speed race tracks. We needed to re-evaluate every item on the TWR XJS.

The initial engine proposal of 400bhp when our programme was started in 1982 was already obsolete. The XJS would need a greater performance increase than had been initially anticipated.

While increased outputs from the Group A specification V12 were possible it would not come without problems. As the engine outputs and rpm had increased in 1982, under-bonnet and oil temperatures rose and this was a problem that needed our attention. The high sump oil temperatures, and double oil coolers in series, meant the internal crankshaft-driven oil pump could barely produce enough oil to maintain oil pressure and lubricate the engine bearings safely. The problem here was that the internal oil pump, driven directly from the nose of the crankshaft, ran at engine speed, meaning the pump speed could not be increased by indirect gearing to increase output. When the car exited slow corners at low rpm, the oil pressure

Charlie Bamber fitting a V12 engine to the Kidlington dynamometer.

warning light, set at 30psi, would sometimes flicker, and although this momentary low pressure had not directly led to an engine failure, it was a distraction the drivers, especially new ones, didn't need.

The project to rectify this problem was one of Jaguar's first contributions to us at EngDiv. in 1983. Tom Seal, from Jaguar Experimental Procurement, sourced 20% wider oil pump gears with the tooth form fully machined. The pump housing was machined deeper to match the thicker gear, thus increasing its pumping capacity, and this fixed our problem.

Another consideration was when using the 97/98 RON regular pump fuel the regulations stipulated. It varied slightly from country to country; Italy especially had a slightly lower grade, and you could never be sure with Czechoslovakia, which was still under Russian control at the time. With our gradual increase to 11.5:1 compression ratio by the end of 1982, and the distributor-controlled ignition curve, we were very wary of detonation, and had subsequently been cautious with maximum ignition advance timings. Conservative (retarded) ignition timings also have the side-effect of increasing exhaust gas temperatures.

Continual work with the mechanical advance setup in the distributor, and the ability to advance the ignition timing 2° after maximum torque rpm, allowed some reduction in exhaust temperatures via these detailed advanced ignition settings.

There were three problems with the standard exhaust manifolds: internal size, design, and heat. Heat from the exhaust manifolds gave the drivers problems when it transferred to the bulkhead close to the driver's left leg. After a while the interior floor in that area had become too hot to touch.

Some manifold variations used on the V12 E-Type and other V12 engined models existed, and we tested them all on our new dyno facility, but they yielded little or no improvement.

In the homologation paper the inlet and outlet sizes of the exhaust manifolds were restricted to the standard measurements. The internals could, however, be hand-finished, and with this in mind each of the exhaust manifolds were cut open on a band saw and the internals completely reworked to assist flow while maintaining the standard entry and exit dimensions. The two parts were then carefully welded together, and the welds ground back to the original surface finish so it could be safely said that no material had been added, as stipulated by the rules. The resulting reduction

Piston design used in the Group A engine. Universal valve cutout design allowed for one piston shape that would suit all cylinders.

in exhaust back-pressure and restriction hotspots was a good move forward, but the greatest power improvement now came from tuning the lengths of the two secondary exhaust pipes. These produced the best result at 26 inches in length before being ducted together into tailpipes 3 inches in diameter. The results from ignition advance and this exhaust development both increased power and reduced exhaust manifold temperature. The original Jaguar exhaust heat shields were retained bolted to the top of the exhaust manifolds, and heat resistant material had been added to the car's bulkhead.

V12 exhaust manifold.

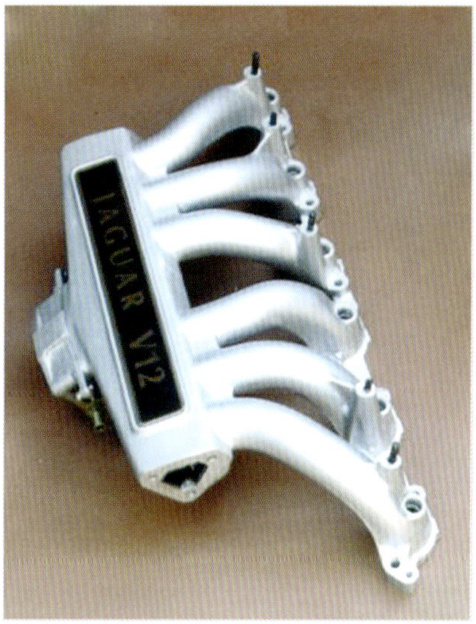

Inlet manifold.

The homologation process mandated the use of standard inlet manifolds too, with intake and outlet diameters restricted to production measurements, and the two inlet manifolds had received this same treatment in 1982. To isolate heat radiated from the top of the engine, we had placed heat shields between the cam covers and the inlet manifolds, and these were upgraded to a sandwich of stainless crinkle sheet material wrapped around a fireproof rock wool inner lining.

Back in 1982 we had provided cool air from around the leading edges of the headlights by cutting holes in the rear of the headlight housing, leading through a flexible connection to the standard engine intake air boxes. Now we improved on this with a three-inch tube fitted inside the air boxes to transfer air directly from the ducting at the back of the headlights to a cast 90° elbow that turned the air directly towards the throttle butterfly. The induction system would run without an air filter, with the original air-filter housing acting as a heat shield for the intake tube. This was necessary because the air intake system was positioned directly above the exhaust manifolds.

One of the problems servicing the race car during pit stops was the strictly limited number of personnel allowed to work on the car during the race – four mechanics plus the refuelling team of two. Refuelling and wheel changing already kept these six fully occupied, and the drivers had to change over within the time needed to complete the tyre and fuel stop. This meant any engine servicing had to be done after two of the four pit crew had finished

Jaguar supplied the information for the erratum process and scrutineering procedure to a brief from EngDiv.

Mr Allan Scott
Tom Walkinshaw Racing Ltd
1 Station Field Industrial Estate
Kidlington
Oxford

Dear Allan

Attached is a draft of the procedure I have requested our Service Dept to have printed. Would you let me know if this doesn't meet your requirements.

Yours sincerely

T N Crisp
Group Chief Engineer
Power Unit & Transmission Engineering
TNC/CW

VALVE LIFT & TIMING CHECKING PROCEDURE

To ensure that the valve lift and valve timing are correct to specification it is recommended that the following procedure is adopted:

Rotate the engine until No 6 inlet valve on the relevant camshaft is fully open. Adjust the tappet clearance on No 1 cylinder to 0.038".

Attach a dial gauge indicator to the cylinder head such that the extension foot is at right angles to, and in contact with the cam follower, and adjust the indicator to read zero.

Rotate the engine in a clockwise direction until the crankshaft position indicator is at T.D.C. and the inlet valve on No 1 cylinder is beginning to open.

The inlet valve lift as measured on the dial gauge should be 0.030".

Continue to rotate the engine until the No 1 inlet valve is fully open when the dial gauge should indicate 0.375".

When the above check has been carried out the tappet clearance should be re-adjusted to 0.012".

With the new specifications, and the experience gained in 1982 with various camshaft periods, I supplied the basic information on profile duration to Cosworth, to produce a profile designated EA3 to the now allowable maximum camshaft lift of 0.425in (10.5mm), using a 74° half-period duration. (Multiply by four to obtain the gross duration after 0.012in tappet clearance; in this instance it is 296° total duration profile.) This process had produced an additional 0.035in of valve lift.

More semi-finished V12 camshafts were supplied by Jaguar, with machining completed except for the camshaft lobes, which were left as cast. The camshafts lobes were formed chilled cast iron, large steel chills being inserted in the casting pattern work adjacent to the lobes to produce the hardening process.

To gain greater definition and durability via a larger nose radius from the new profile, a larger 0.600 inch base circle radius was chosen, producing in turn a larger camshaft lobe. However, the larger lobe would not now fit within the confines of the existing cam carrier and cam cover, and the entire cam carrier needed to be raised to allow for the larger base circle radius and increased valve lift. The new height would allow sufficient tappet clearance when using the standard valve lengths. The camshaft timing was set to maximum inlet valve lift at 102° after top dead centre (ATDC).

Under the ruling that stated gaskets were 'free', an aluminium spacer gasket was placed between the upper head surface and the underside of the cam carrier. This raised the cam carrier and camshaft centres, allowing the new larger camshaft lobe to be used.

The cam carrier housing was relieved in the areas adjacent to the camshaft lobe, and allowed the larger lobe profile to rotate freely. The timing chain had two extra links added, which brought the tensioner system back to its normal operating range with the raised camshaft centres.

We had also experienced some fatigue failures where the V12 chain tensioner spring had broken at one of the hook ends, and it was determined that a small crease in the mass-production process was the cause of the problem. Jaguar was commissioned to rectify the problem and once again Tom Seal came up with a very successful fix, employing solid eyelets at each end with a new spring wound around them. EngDiv. continued to use this system in all later V12 engines including Group C.

EN or Emergency Numbers to identify different types of steel were introduced during WW2. In general terms the lower the number the lower the carbon content.

The letter of the alphabet after the number denoted the heat treatment process after machining.

General purpose steels	1–16
High quality steels	19–24
Hardwearing, e.g. camshaft	36
Spring Steels e.g. crankshaft	40
Stainless Steels	56–60

America uses its own different numbering system.

The cam followers that actuated the valves were increased in diameter from 1.375 to 1.4 inches and made from hardened EN 24 steel. The larger cam follower diameter allowed for a higher valve acceleration to be used in the new camshaft profile design without over-reaching the follower's outer diameter. Using smaller diameter followers with such a design would introduce potentially damaging side-thrust to the follower bores as the camshaft's initial lifting took place on the outer edge. The high valve acceleration rate in the early lifting zone of the profile helped to mitigate some of the performance losses when restricted to an overall lower-than-optimum maximum valve lift.

From mid-1982 George Toth at Richard Longman and Co. in Christchurch, Dorset, had carried out the gas flow development work on the V12 heads and inlet manifolds, and this had produced good improvements. When cut open the inlet manifold dimensions to each cylinder were now looked at in detail to produce a more uniform air flow to the cylinders, this to improve the conformity in cylinder to cylinder air-fuel mixture, which had varied slightly.

New valves were manufactured to a design for better air flow, but maintaining all the standard dimensions.

The standard throttle butterfly fitted to each of the two inlet manifolds sits in quite a deep housing that keeps the initial butterfly openings shrouded and leads initially to a non-progressive air flow against accelerator pedal movement. A machined taper was introduced in the housing on either side of the throttle butterfly, merging at the closed butterfly points to maintain the homologated throttle size. This modification increased airflow at lower throttle openings, resulting in a sharper initial throttle response and a more predictable pedal-to-power movement through initial part-throttle openings.

The standard EN 16 high cutting speed material head nuts, needed for mass production had shown problems with thread quality because of the material's sulphur content, but with the help of the Jaguar's Prototype Department we introduced new domed nuts, having improved material and heat-treatment of EN 16T. These nuts had to be specially made as they needed a domed top closing in one end to stop any water that might come up the threads from the studs exposed to water circulating in the block. New head studs were also made to standard specifications, but with the original EN 16 material replaced with EN 24W.

All these modifications increased the power output as well as adding increased engine reliability. By the start of 1983 power had been increased to a maximum of 420bhp at 7000rpm, with only a slight drop off to 410-415bhp at 7250rpm. Maximum torque of 365lb/ft was produced at 5000rpm.

We had also achieved maximum brake mean effective pressure (BMEP) of 175psi, or 12 bars. This is a measure of how well each cylinder is filling with air, and the mechanical efficiency with which combustion produces pressure on the piston during the combustion cycle. It is an important measure of an engine's mechanical and thermal efficiency. The standard V12 engine has a BMEP of 141psi, but the open exhaust Group A engine produced an improvement of almost 25%, still using modified standard components.

ENGINE TEST

DATE

ENGINE TYPE V12 SPEC 82 pistons 11.6:1
ENO. T.W. VALALUNGA. EA3 (10°)

LOAD

R.P.M.	1	2	3	4	5
4000	61.5				
4500	67.5				
5000	70				
5500	68				
6000	65				
6500	62				
7000	60				
7250	56.5				
7500					

	B.H.P.	Torque
4000	I.A. 246	323
4500	304	354.4
5000	350	367
5500	374	357
6000	390	342
6500	403	325
7000	420	315
7250	411	296.6.
7500		

"V12G1AJAG2"
Disc 10

Full load Oil pressure at
Oil Temp

Typical power curve at the start of 1983, this engine for the second race at Vallelunga. Power by year's end increased to 430bhp. TW would normally restate all engine programme power figures, reducing them by 5% in his endeavours to show a competitive TWR race car package. Most teams usually slightly exaggerated these power outputs.

The ERA analogue system that had been fitted from the start of 1983 entailed a steep learning curve, and although a large step forward, there were some teething problems. The analogue fuel settings had to be programmed in a set sequence, with slopes having to be followed by a flat section before another slope could be programmed. As detail changes were made to the engine specifications, so the amounts of fuel needed changing as well, and this programming process could only be done by Bill Gibson at ERA.

Analogue fuel system

We can compare this analogue control system to that of the mechanical injection systems used in most racing engines of the 70s and early 80s. The second graph shows the electronic throttle angle signal opening slope, and duplicates the mechanical shape produced by the throttle cam in a mechanical injection system. There is little difference between these electronic and mechanical controls as the engine advances from closed to full throttle. The difference comes with the graph below with the fuel amounts varying and tailored to the engine volumetric demands.

A mechanical system can only have a constant injection output, and would be a straight-line somewhere in the upper mid-point between 6.00 and 7.25. With mechanical injection, the fuelling would be set to give best results between maximum torque and maximum horsepower, leaving the fuel mixture to run rich at both lower and higher rpm and slightly leaner at maximum torque. These problems were now corrected by tailoring of the electronic injection system producing a slightly wider power band and improve fuel consumption.

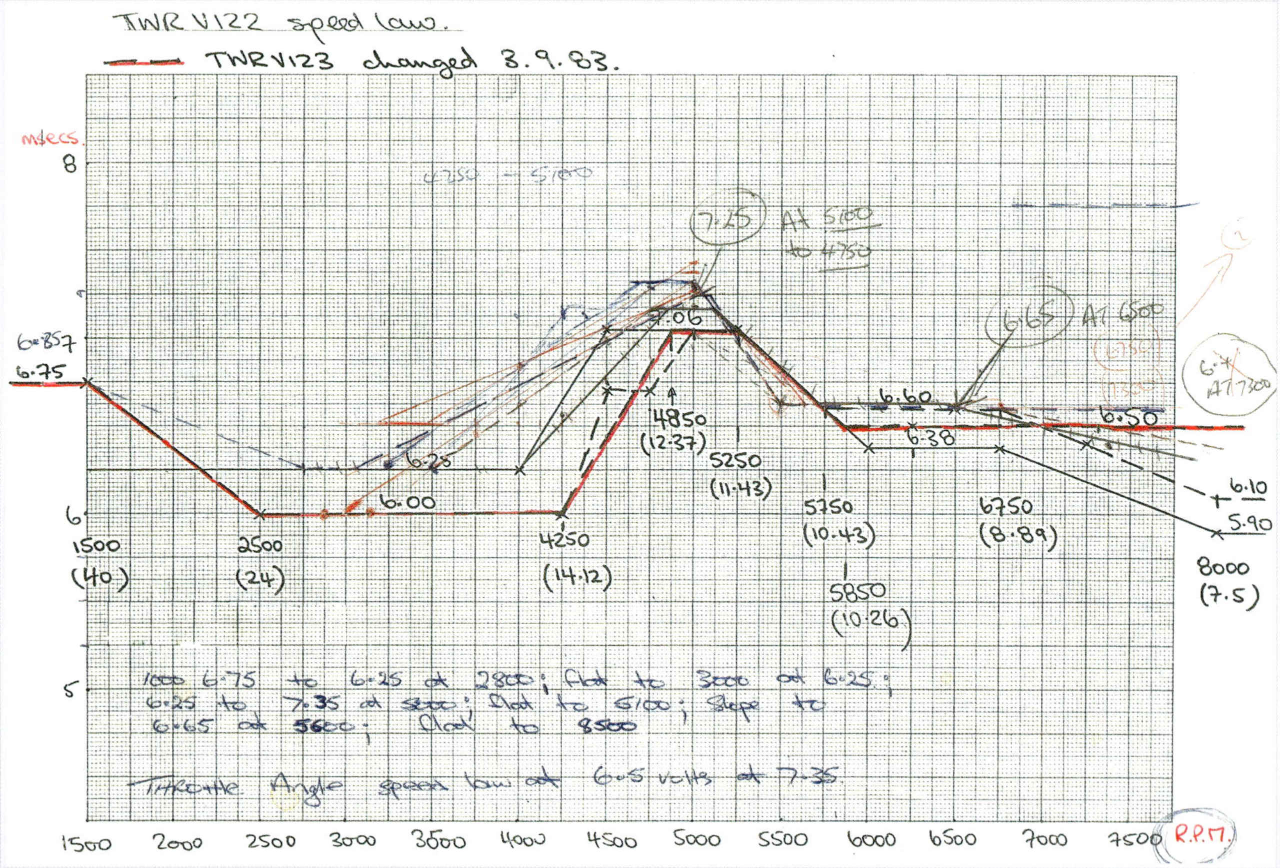

Fuel demand at full throttle showing the fine tuning carried out during the year.

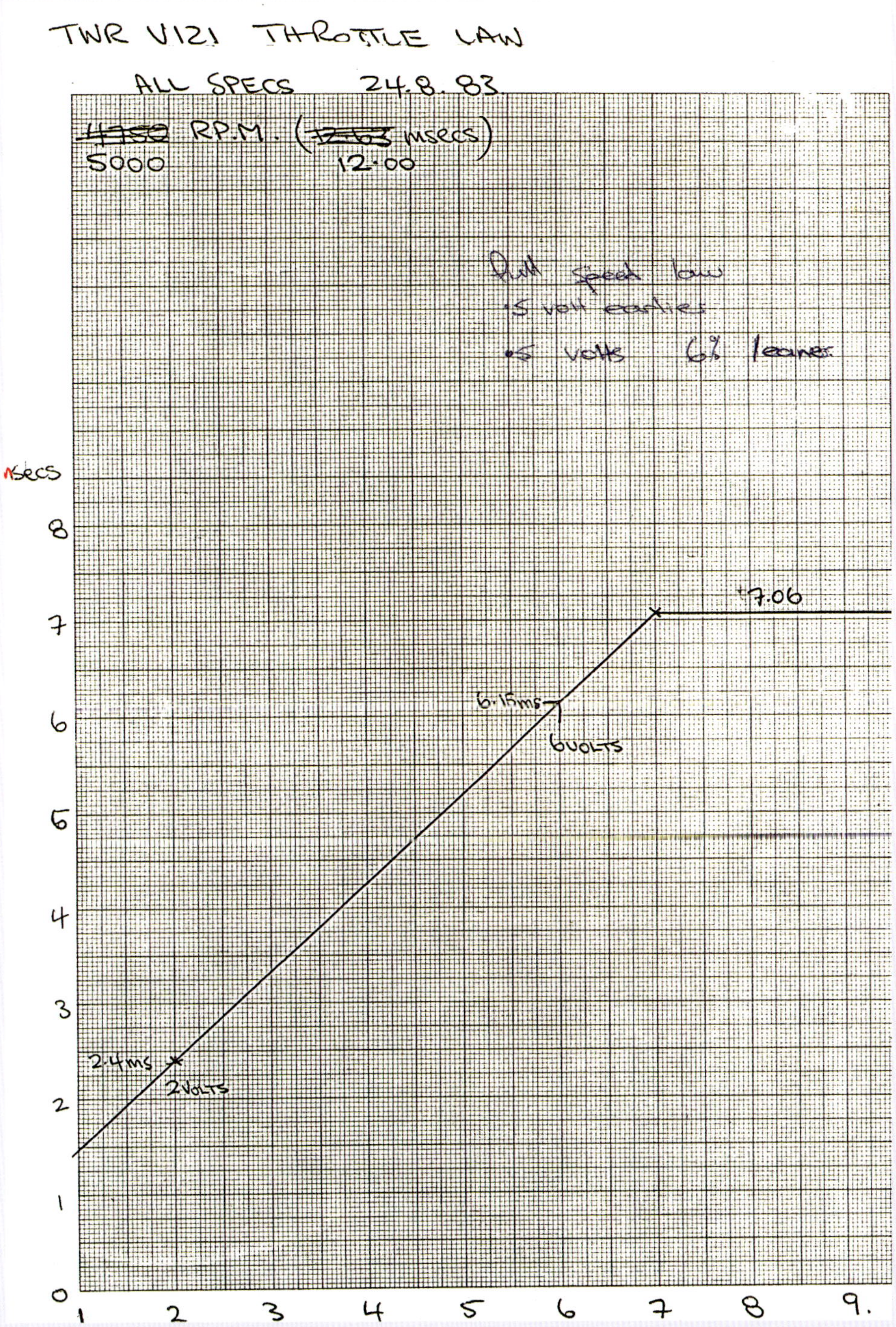

Best compromise was set at 5000rpm for throttle response after gear changes.

Throttle angle slope shows full fueling before wide open throttle signal at position 8.

The graphs (on the previous pages) show the many computations tried during the year, with the many variations in detail needed as the engine approached maximum torque between 3000-5000rpm. It also illustrates the changes in fuel requirements after 6500rpm, with the various tuning methods having their effect on fuel demand as volumetric efficiency decreased.

The XJS chassis incorporated all the lessons learned in our crucial first year, and updates were added over the winter. Brake disc diameters and thickness had been nominated at the start of 1982, before testing could be carried out, to allow homologation. These items now needed to be fine-tuned to achieve more acceptable temperatures and mechanical braking balance front to rear. Brake calliper and master cylinder piston sizes needed to be addressed to enable them to work together and not cause problems with articulation in the adjustable brake balance box situated between the brake pedal and the master cylinders. Brake ducts for air cooling had also been a constant problem in 1982, especially around the front wheel areas where ducts could easily foul and detach as the wheels were turned on full lock, causing loss of air cooling and leading to overheated brake discs.

This was way before the days of modern carbon or ceramic brake pads. Overheating DS11 disc pads wore unevenly because of the temperature differences across the pad surfaces top to bottom, leading to a wedging or tapered wear effect. In turn, this resulted in a rocking motion as the tapered pad was applied to the flat disc face, meaning the driver had to apply more brake pedal movement to engage the pad with the disc. This effect is called a 'long pedal'. DS11 pads also required a high pedal pressure, especially when cold, because of their low coefficient of friction, but fitting small master cylinder sizes to increase hydraulic pressure exacerbated the long pedal problem later in the race. Fine-tuning of calliper piston sizes, disc thickness and cooling vane types and front and rear master cylinder sizes were all allowed under the rules, but some of these items had to be recognised on the FIA homologation papers.

Front brake disc and suspension assembly. Note open face calliper for quick pad changes.

New designs with different internal cooling vanes and disc widths, from brake supplier AP Racing, were constantly being tested and had to then be approved. The front brakes were also a constant problem because the 13 inch diameter discs were still too small for this heavy car on high-speed circuits. However, that was the largest size that could fit within the confines of the 16 inch diameter wheels being used.

Rear disc brakes with ducting.

The XJS was slightly more front-heavy than the average car [56%] because of its relatively short wheelbase and the heavy V12 engine's forward mounting, meaning that the race car could not carry increased rear wheel braking to relieve some of the pressure on the front brakes. This resulted in a very high workload on the front brakes when slowing almost 1600kg of driver, fuel and race car. The team experimented and refined a water injection system to the front discs, applied via a windscreen washer electric motor each time the brake pedal was pressed, and this helped control the problem as the 1983 race season progressed.

Outboard rear brakes on modified hub castings had been homologated in 1982, and the original hubs, with welded mountings for the brake callipers, were now replaced with the new castings. These changes all took time, and kept Ron Elkins busy with the homologation paperwork at Jaguar.

The 1983 cars had the gearbox oil coolers relocated under the front windscreen in the heater plenum area which produced a much tidier arrangement.

Jaguar Cars were experimenting with weight reductions on their future models using thinner glass and attention to detail with the new TWR003 car and the acid dipped TWR002 body shells, now trimmed the weight down to 35kg above the 1400kg limit.

The motor racing situations at TWR and its increasing complexities with the need to win, be financially rewarding, and take control of two manufacturer's racing programmes that looked to have an increasing future in the sport, did not present an easy task to anticipate or control in 1983.

Gearbox oil cooler taking air from beneath the front windscreen.

The FIA 'yellow book' of rules was by now a close companion, well thumbed through in the Group A section. De-regulated items in the rules normally used the word 'free', which could open a can of worms for the regulators. If nuts and bolts were 'free', could they also be used in another area to perform a different function? Could it be argued that something else performed the function of a nut or bolt, and was therefore 'free'?

The rules had first been written in French, and then officially translated into English, and TW had considered the use of a Queen's Council to give a legal opinion on the meaning of certain words without being influenced by preconceived engineering beliefs. One example I had found with the Rover V8 programme was that the Oxford Dictionary's definition of a tappet included far more of the valve train than we all accepted, and this had encouraged us to often apply a wide and free interpretation when reading the rule book.

Such semantics were instinctive to TW; he excelled with them and we all added our thoughts with some enthusiasm. It was a lesson in thinking outside the square and focused the team on what might be possible rather than the obvious and perhaps traditional path. Some of his less practical ideas though just had to be given time to allow them to move lower down the priority list, and fade away.

TW used to say there was no black or white, only shades of grey, in this closely regulated type of motor racing for manufacturers in Group A. This had become difficult for racing enthusiasts to accept, but it was an increasing fact of life in this competitive arena. He also knew that a manufacturer would not last long in the sport if all they did was just make other manufacturers look good, by them constantly filling the minor places.

The new Group A Jaguar workshop.

In 1983 there would be just two manufacturers competing for outright honours in Group A. Two years later however, Jaguar and BMW would be joined by Rover, Volvo, Ford, Mercedes and sometimes Toyota and Mitsubishi, such was the fast developing nature of this increasingly successful series. The days of the heroic 'privateer' entrant in the UK were being challenged and motor racing had to move with the times. If the manufacturers with money were interested in marketing their products, they naturally needed to win. The more of them who entered the sport, the harder it would become to win. Taking a manufacturer who wanted to be in motorsport, but with their vehicle designed without that concept in mind was a whole new ball game, and the XJS fitted into this category.

Existing race teams were generally quite small, and some had become dominant over the years but were now being challenged by increased competition and professionalism with the advent of Group A. Thinking outside their proverbial square also cost money, something they might not have.

Jaguar in Coventry also supported an American racing team called Group 44, but not yet at official works level. Jaguar management was already thinking of an increased involvement for that company in future motor sport plans, and Group 44 was to fill the role in Jaguar's USA expansion as well as thoughts of a return to Le Mans. Later in the year, when we first attended meetings with Group 44 team members, I observed that team owner Bob Tullius was much more cautious. His one goal seemed to be reliability and finishing. Better to finish and get a place than to try something new for an outright win, seemed to be the philosophy at Group 44. It was obvious to me that the two Jaguar-sanctioned racing teams, Group 44 and TWR, saw success at two different levels. At TWR you either won or lost; there was only one step to be standing on. The second step was for the first loser and Jaguar was going to have some difficulties ahead in trying to have these teams work together. Something we had been reminded of by them in our meetings.

The XJS needed more aerodynamic downforce at the rear, with long fast corners on European circuits and high-speed braking taking its toll on the rear tyres and front brakes. There was huge weight transfer to the front under braking, and the XJS needed its rear end held down, both to aid stability and produce good grip in the long fast corners.

BMW had fixed their similar though lesser problem by fitting a boot spoiler to the 1983 Group A 635 CSi coupe. Roof lines and rear window angles were being selected by production car aerodynamicists to help the airflow bend down towards the trailing edge of the boot lid. This engineered air flow helped the lip spoilers and aerodynamic devices on the trailing edges at the

The controversial rear 'oil cooler' for 1983.

rear of the vehicle to reduce lift, and even produced small but vital amounts of downward force at high speed. The XJS's design pre-dated many of these concepts, as it had already been in production for eight years and its basic design principles originated in the late 1960s. As a result the XJS race car had no appendages to assist in the new aerodynamic trend for reducing both lift and drag.

That problem led to the creation of the so-called 'oil cooler', which now emerged from the rear of the XJS at the start of the 1983 season. Fluid coolers were 'free', as were their mountings, and the overheating differential oil certainly needed cooling. A full-width single tube differential oil cooler was fitted low underneath the rear bumper, along with a suitably angled 'mounting bracket'. Regulations permitted the use of oil coolers with intakes situated under the car. To most observers though, it looked like an under slung aerodynamic device, or wing, and they were probably right to think that. To TW though, it was very obviously an oil cooler with a large angled mounting bracket. This device was straight from the Tom Walkinshaw 'outside the square' book of aerodynamic and engineering solutions.

Monza – Italy

It was March once again, and that could only mean one thing – Monza in Italy.

The two cars were carried to Italy facing forward in the transporter and as the tail-mounted lift was dropped on arrival in the Monza paddock, it revealed the new rear oil cooler fitted to both cars. Helmut 'The Fox' Kelleners, BMW's most senior driver, walked over for a look. He shook his head. 'Tom,' he exclaimed, 'you have come too early for the Group C sports car race, it is Group A racing this weekend …'

The idea behind it was that the handling benefits and longer rear tyre life would more than compensate for the additional drag of the 'cooler'. The other option without this device was being able to overtake on the straight because of less drag, while being a little slower in the corners transferred the problem to the car behind. That would be an interesting point for the rest of the season; lap time versus track position during the race.

It was also a case of TW wanting to be seen to out-think the opposition, and move some of the motor racing off the track; he excelled at that. Discussion of the subject would also produce more 'free ink', which was probably better for TW than Jaguar. As far as I was concerned this 'oil cooler' had little going for it.

The important thing in developing the regulations as far as TW was concerned was that competitors had to listen to his interpretation of the current rules, and the opposition were not as good at presenting their case against his interpretations as he was at presenting his version. It was a path that required some discipline and judgement however, to ensure it was to not become the complete focus of the team's thinking. Individual chassis engineering expertise still needed to be developed and encouraged within the team, but that was not easy with such a dominant character at the top. On the other hand he was happy to leave the responsibility of the engine programme to me, and that suited me too.

Helmut Kelleners.

BMW turned up in force at Monza. It had 10 new 635 CSi coupes in the hands of teams representing a range of European countries, far more than the small number of 528i saloons represented the year before. Previously, fewer new coupes had been manufactured than the 5000 per year that was needed under the regulations, but production had been increased to help their hasty homologation for 1983. The BMW's minimum weight was 1185kg, and its 3.5 litre engine was capable of power outputs in excess of 280bhp, a combination that produced good handling and fuel consumption. Some of the cars were even equipped with Formula 1-technology Pirelli tyres.

Even though the 1983 XJS at Monza was a far cry from the car that had arrived 12 months previously, the new 635i BMW was equally a large step forward from the 528i. The 1983 XJS design concept started at Monza with a best minimum weight of 1435kg, and 425bhp, giving it a power advantage. The 635i had a better body aerodynamic design, with a lower drag coefficient to help

From left: Chuck Nicholson, Tom Walkinshaw, Pierre Dieudonne, Enzo Calderari.

make up for its lower power output, an asset on Europe's fast race tracks. Both cars had similar brake sizes because they used the same wheel diameter, giving the lighter BMW an advantage in braking and cornering.

With BMW's large step forward in performance and increased focus on the ETCC, some old battle lines were being redrawn in modern European Union sand.

The XJSs of Walkinshaw and Pierre Dieudonne, however, were on the front row after qualifying on Dunlop's soft 922 compounds. The Jaguars were three seconds ahead of the first Schnitzer BMW, driven by Hans Stuck and Walter Brun, followed by the similar Eggenberger cars of Kelleners and Grano. Impressively the Jaguars had qualified almost 10 seconds faster than what had been achieved in the first race at Monza just 12 months before, clearly indicating the progress with car development and tyres over that period of time.

On the other hand, BMW's switch to the 635i brought them an amazing seven-second reduction in lap times – all in just one year; one to two seconds per year being more the norm in this type of saloon car racing. This huge step forward showed Group A was now being taken seriously by competitors and manufacturers; engineering solutions applied to current saloon cars were already probably taking the series to a higher level than envisaged when the rules had first been applied just over a year earlier. In 1983 there were 39 starters for the 87-lap, 313-mile race. It showed the coming of age of the second year of Group A, with BMW being represented by no fewer than 10 cars.

Jaguar senior management look on during the first race at Monza. *Left to right:* John Egan, Jim Randle, Neil Johnson.

From 5000 in 1982, the number of spectators at Monza had increased to 12,000 in 1983. Group A was turning into a good business for circuit promoters and much of this new success was coming from TWR's involvement.

From the race start the Jaguars had things under control until one-third distance, when the Dieudonne/Calderari XJS encountered engine overheating and failure; its race was finished. After the last pit stops, and two-thirds distance, the Walkinshaw/Nicholson Jaguar was a minute in front, when the bonnet retaining pin at the trailing edge on the driver's side ripped from its mounting. The raised bonnet obscured Nicholson's view and forced a stop for repairs. A hastily made tape bandage was readied before the car was called in. However this hasty repair was not very effective and Chuck had to continue with some impaired vision.

It was one of the most spirited drives we would ever see from him, as he battled his way back from fourth place to finish within three and a half seconds of a victorious Quester, who crossed the finish line in a series of huge controlled power slides to celebrate top German preparation team Schnitzer's first win for the BMW 635i.

During later wind tunnel testing at Mira it was found that the XJS could not successfully vent the air from under the bonnet after it was pushed through the radiator at high speed. The inability to easily vent this air had led to increased under-bonnet air pressures

Nicholson with the hastily repaired bonnet.

The finishing orders for the first race of 1983 were:

Quester/Rossi in BMW 635i
Walkinshaw/Nicholson
Hans Heyer/Armin Hanne
and Thierry Boutsen/Bruno Giacomelli in BMW 635is

at the constant 160-165mph speeds achieved on the Monza straight, and eventually causing failure of the retaining pin fixing in the body.

Detailed modifications at the rear of the bonnet were made after this race to help the air escape. Body return seams were turned back and the number of retaining catches was increased to four, set so that the rear of the bonnet would lift slightly and vent as the air pressure built up underneath. This allowed some of the air to flow through the engine bay, and some over the top of the car instead of under it. This slightly reduced the aerodynamic lift and improved the radiator efficiency.

Vallelunga – Italy

Vallelunga, near Rome, was a tight track that suited the nimble BMWs. The Jaguars managed second and third fastest times in qualifying amongst the 32 starters for the 157-lap, 310-mile race.

Walkinshaw led the first part of the race until one of his three-piece wheels broke, and although he was able to get back to the pits and have a replacement fitted, he returned to the race well down the order. The Nicholson/Dieudonne Jaguar regained fourth place after its pit stop, and Calderari later took over from Walkinshaw, replacing Nicholson.

During his stint, Calderari was punted off the circuit by Manfred Winkelhock in a move Enzo adamantly claimed was deliberate. It was not unusual practice to bring in special drivers from other race series that could be called on occasionally when commitments allowed. These ring-ins could serve another purpose, being able to take a different short-term view because they were not competing in the full championship. If you were a full time competitor in the Group A series, and carried out a deliberate hit such as that on Enzo Calderari, one day it would be your turn to receive the same treatment – perhaps at a very important stage in the series when every point counted. Generally, the professional drivers behaved very well over the relatively long races, all knowing that if they didn't there would be some payback later.

Tom Walkinshaw takes a closer look at the modifications used on the BMW 635i.

The race was won by Kelleners/Grano, followed by Quester/Winkelhock, with Walkinshaw switching to the Nicholson/Dieudonne Jaguar and coming home third; but he was unable to gain points under the rules regarding car-switching during the race. The next six places were taken by BMW in professional teams run by Schnitzer, Hartge, BMW Italia, Juma and Lucky Strike.

John Goss and author Allan Scott in conversation at Vallelunga.

It was beginning to look a little lonely for the two-car TWR Jaguar team when one of the cars failed during the race. 1983 was going to be a much more intense year and vehicle eligibility whispers had started already.

Australian John Goss had attended this race as well while on a visit to the UK, and had been to Jaguar for assistance with his XJS entered in the Bathurst 1000km race.

He had come to Vallelunga to observe the TWR operation in action and hoped to raise increased interest in his attempts at winning this high profile race held each year in Australia. The Australian rules however were quite different to FIA Group A and few parts were interchangeable. I had been assigned the job of 'liaison and interpreter' to keep a watching brief on this outcome as the seeds of venturing to Australia for this race had already been sown when TW and Elizabeth had been invited to dinner with us one evening in 1981 to watch the latest Bathurst tape I had been sent.

Unknown to Goss at the time I had already been in contact with CAMS CEO J. Keeffe in January with the prospect of the Australian rules aligning with FIA Group A.

Donington – England

Donington was a welcome break for the long distance travelling TWR team, being just a few hours' commute from Kidlington. It was a large team that turned up for this race with the BSCC TWR Rovers, now built to Group A specifications, contesting their first ETCC round.

It had been noted by Jaguar executives in 1982 that TW's own private car was a Mercedes-Benz estate, and a slightly embarrassed Jaguar boss made a note to provide him with a Jaguar from the car pool at Browns Lane for next year's ETCC round at this Midlands circuit. However, they made the mistake of lending him an

watering of the eyes, I remembered the old ANZAC code, kept my head, said nothing and returned a weak smile. The war was over, for another day.

There had however been a logistics problem at Donington, with not enough correct tyres available for the wet conditions. TW had the better new 54 compound, while John Fitzpatrick had been allocated the less competitive 622 tyres to start the race. Fitzpatrick appeared to struggle at the start of his stint, while TW carried a faster pace in the early stages of the race. When it became apparent that the Calderari/Fitzpatrick/Brundle Jaguar was in contention for a win after the refuelling mishap with the leading Jaguar, the new softer 54 compound tyres were fitted to that car for its last stint, but obvious favouritism had been shown to the team boss. Although not a lot was said at the time, it had created a potential problem between the pair and while John Fitzpatrick had been the senior driver in their previous time together during the mid-1970s, these roles were now reversed.

Enna Pergusa – Sicily

The second victory for the TWR Jaguar works team came after the long trip to Sicily, at the Enna Pergusa race track. Thirty-six cars started the race, with Walkinshaw taking the pole time in qualifying. Walkinshaw and Nicholson went on to win by one minute from the second-placed BMW of Kelleners/Grano, with Vanoli/Hollinger next and Stuck/Brun fourth. It was a difficult weekend because of the heat, both the Jaguars suffering temperature problems throughout, and the Calderari/Dieudonne Jaguar failed to finish the race because of engine overheating, after qualifying in fourth place.

The championship standings were now Quester on 74 points, and Walkinshaw on 60 in second place.

Paul Davis talks to Chuck Nicholson.

As the ETCC programme continued, Jaguar's future plans for motor racing were a constant point of discussion at the monthly meetings at Browns Lane. While the racing fraternity watched the TWR XJS team battle to have supremacy over its BMW opposition and assumed all its resources were focused on this outcome, the following notes of the May 1983 meetings held at Jaguar show the diversion of resources and forward planning for Jaguar's race programmes. Already the move to sports car racing in Group C is on the agenda.

RACING PROGRAMME – MONTHLY STATUS – 10 MAY 83

Present		
	Mr J N Randle Mr T N Crisp Mr R M Akroyd Mr D B Birch Mr C G Edwards	JAGUAR
	Mr T Walkinshaw Mr A Scott Mr P Davis	T W Racing

GROUP 'C' ENGINE

Contract Status – Contract being typed up at present. Expected completion and despatch today. Includes staged payments on monthly basis.	D. Birch.
V12 Engine Development – 4 valve heads have been gas flowed Engine capacity TBA	T Walkinshaw
Component sourcing and specification to be rationalised.	R M Akroyd
Specifications to be clearly defined. e.g. crank pins.	T N Crisp
No parts are to be ordered without TWR consulation.	R M Akroyd
Matrix of components against users to be drawn up.	R M Akroyd
Installation drawing of V12 engine in the group 44 car to be supplied.	R Tullius
Heads casting patterns will have to be new. 4 valve casting drawings to be discussed. Lead time requires early decision.	R M Akroyd
Magnesium – alloy cylinder block – 6 off machined. Hold off line boring main bearings until decision made on main bearing size	R M Akroyd
Vandervell are to be consulted.	T N Crisp
1st Engine specification to be clearly defined so that a positive datum is established.	A Scott
Report on V12 4 valve testing at Jaguar to be supplied.	T N Crisp
Decision on chassis for Le Mans needs to be taken see later notes	J N Randle

LUCAS - MICOS DEVELOPMENT

Ignition system has run and tested in Group A car.

Jaguar Power Unit design engineer to attend Lucas meeting — T N Crisp

Date of meetings TBA — A Scott

GROUP 'A' STATUS

Pits work flow - Process analysis and work study personnel available at Jaguar if required. — T Walkinshaw

Cooling spec - Radiator design and configuration on development. — A Scott

Exhaust manifolds - Less constricting design could be used with the additional space available. Manifold must be cast but could be fashioned on fabricate design, already tested. — T N Crisp

Also Series III E Type to be investigated. — T Walkinshaw

Limited slip differential - New type being developed by Ex GKN Engineer. Investigation required. — T N Crisp

Wet weather tyres - CAD to be used for wheel envelope with different wheel/tyre combinations. — T Walkinshaw

Availability of Michelin 410 TRX tyres.

Front 11.5 Rear 12.0 Rim dia. 16.3 used on WM sports car at Le Mans. — J N Randle

Rear spoiler - Considerable advantage found by using mocked up type. Report and sample to be supplied to enable wind tunnel exercises. — T Walkinshaw

Styling exercise currently ongoing should reflect results of testing — AJC/DT

IMS STATUS

'Ball Park' prices for the car to be made and supplied to Jaguar. — R Tullius

Status of 2nd car required and assessment made. Feasibility of shipping 2nd car to UK for 2 day trials at Silverstone, timed early June, to combine with Mr Tullius visit. — T Wilkinshaw

Discussions with Mr Dale to ratify. — J N Randle

Mr L Foushee to be invited to attend next meeting. — J N Randle

4. FORWARD PLANS

'5 year plan' of Racing Programme to formulate common objectives.	J N Randle
AJ6 engine to be included in future discussions.	J N Randle
XJ41 to be checked for compliancce in Group 'A'	J N Randle
Assault on Le Mans - Development programme, and fact finding trip to circuit to measure bend radii etc, for P Lings to develop computer programme.	J N Randle
Review possibility of hiring say Nardo for testing.	T Walkinshaw

C G Edwards
25 May 83
CGE/SD

Next Meeting 5.00 p.m. 7 June 1983 in Mr J N Randle's office.

Circulation

Those present + Mr M Dale
Mr R Tullius
Mr L Fousee
Mr D B Scholes

c/i Mr J L Egan

'Yes Tom, but we did beat you here.' TW and Umberto Grano.

Mugello – Italy

Mugello in Italy was never a kind circuit for the team and my heavy work load meant that this was a race I did not attend. Walkinshaw was once again on pole, with the Kelleners BMW splitting the two Jaguars. Both Jaguars were consuming more fuel than normal, and Walkinshaw's car was struggling with its tyres at a circuit where the team had very little testing time. TWR was not the dominant force here it had been in previous races.

Chuck Nicholson partnered Enzo Calderari in the second car but it was suffering from a mysterious understeer problem throughout practice, and then suffered a rear puncture during the race. The competitiveness of the top three BMWs had improved, with new Pirelli tyres suiting the circuit while the TWR team was suffering from constant long distance travel leaving little time to prepare the race cars. The Dunlop-shod BMWs were off the pace as well as the Jaguars. The winners were Kelleners/Grano, with Vojtech/Enge second and Walkinshaw/Fitzpatrick third.

Brno – Czechoslovakia

Brno in Czechoslovakia once again rewarded TWR with a victory. Both Jaguars showed superior speed, with the second car driven by Pierre Dieudonne circulating closely behind Walkinshaw's leading XJS until a few laps from the end when a broken rear suspension rose joint forced it down to sixth place. The CSSR/Hartge 635i BMW of local Czech drivers Zdenek Vojtech and Bretislav Enge had been the leading BMW team for much of the race until tyre failure resulted in an accident. These two would continue to make an impact on the Group A scene. Quester/Heyer came in second, one lap down, showing the dominance of the Jaguar at Brno. Delcourt/Swyssen were third, followed by Stuck/Brun, with Nussbaumer/Hartge fifth.

The 160mph blind entry into the village of Bosonohy on the Brno street circuit. Turning points were taken from the different coloured front doors of two of the houses.

ABOVE
Left: The Hartge prepared Vojtech/Enge BMW 635i.
Right: Bretislav Enge and Zdenek Vojtech.

For the first time in the season, the TWR Jaguar XJS of Tom Walkinshaw took the lead in the championship from Dieter Quester.

Österreichring – Austria

The fast sweeping corners and long straights of the Österreichring in Austria saw Martin Brundle return to the team for a second time; changing from his Formula 3 RALT single-seater for the drive alongside Walkinshaw this time. Dieudonne/Calderari were in the second XJS.

The Jaguars set the fastest qualifying times, and their only weakness on that track was braking at the 200 metre board after reaching a top speed of almost 165mph on the straight. If the Jaguars could gain a small breathing space on the opposition down the long straight, the cars could then brake earlier and avoid overheating the front brakes.

Time could not be wasted by the crew during the pit stops either because the TWR team had to carry out an extra tyre and fuel stop as usual, and although the Jaguars were running two seconds faster than their BMW opposition, the margin at the end of the race would be fine. A 20-second pit stop still meant 50 seconds out of the race including pit lane entry and exit at high speed, fortunately there were no pit lane speed limits as is normal today.

Paddock whispers had grown to the point where BMW lodged a protest against the TWR Jaguar's eligibility, with BMW even producing their own new XJS, parked at the rear of the pits, as a

Quester and Stuck, what are we going to do about …

reference vehicle in support of their arguments. BMW boss Dieter Stappert was on hand to sort the various BMW teams' problems with TWR, starting with the rear suspension and then engine capacity. As had happened in the UK with the Rovers in 1981, I had to be happy when my peers, in this case BMW, thought the successes with the TWR race cars could only have been achieved with a larger-capacity engine; perhaps it showed that EngDiv. were doing things better with production engines than the Germans thought possible. The rear suspension protest was about our moving the inner wheel arch panels to fit the recommended wheel size, as well as produce a ruling on the rear 'oil cooler'. TW counter proposed that we would accept the inspection provided the BMWs were inspected as well, and his V12 engine was then sealed by the stewards.

Officials overnight declared all was well but Stappert was not happy. His gamesmanship continued the following day when he showed race officials a report indicating the TWR pit crew had five people working on the car as well as the re-fuelers, when only four were allowed. The officials agreed that this was outside the rules and then asked him what they should do with their report that showed that seven mechanics had been seen working on one of the BMWs during the race!

At one point during this race Martin Brundle had to deal with a slowly-driven BMW deliberately pulling over in front of him, and it was a small miracle that his Jaguar was not put out of the race after he valiantly managed a wild slide while avoiding contact. The dangerous manoeuvre was carried out by the same driver who had pushed Calderari off at Vallelunga. He was co-driving with Quester in the same car which had been held up on the wet Donington track where the team felt they should have won. Those antics were part of the sport though, were they not? Amazingly these manoeuvres continued once or twice more, but at least Martin was now ready to take the necessary avoiding action.

The event finished with rain during the last few laps, and both Walkinshaw and Dieudonne drove the final well-controlled laps to keep the 'rainmeister', Hans Stuck in his BMW 635i, at bay. It was the first 1-2 for the TWR team in 1983, but was finished

From left: Brun, Stuck, Brundle, Walkinshaw, Calderrari, Dieudonne – Österreichring.

off with an extensive protest which kept the team at the circuit until midnight.

BMW was not happy with the recently-announced large percentage increase in sales of Jaguar vehicles in Germany, albeit from a very low starting point. After much discussion, the cars were cleared by the stewards and we were on our way back to the UK aboard a private jet. Chartering aircraft to cover racing events was now becoming increasingly necessary as TWR travel needs and tight deadlines could not always be met by scheduled commercial flights, unlike the advantage of the European teams who were just a few hours from home at high speed on the autobahn.

Nürburgring – Germany

Next it was back to the daunting Nürburgring in the BMW heartland – Germany, scene of last year's Jaguar win. This year though, there were five TWR race cars to manage – the normal Jaguar programme and three TWR Rovers from the BSCC sprint race series. We were all working late nights, and I had missed the earlier Mugello and Brno races to concentrate on engine rebuilds. The EngDiv. staff had grown to five with the arrival of Peter Grace and then Dave Morris, both former employees at Hesketh engines. Dave Morris would later go on to be one of Ayrton Senna's Formula 1 mechanics at Lotus.

It had been decided at relatively short notice to take the TWR V8 Rovers to Germany, even though these BSCC Group A

Rebuilding the V12 engine in less than ideal conditions overnight.

Around mid-morning the zip was pulled down on the awning and two heads poked through, to the consternation of stressed and tired mechanic Dave Roberts. With his arms flailing, Dave told the two heads to 'f... off!'

The heads replied that it was: 'OK: they were from Jaguar.'

Dave's reply was a classic: 'I don't care who you are, even if you are John Egan, f... off.' After the outburst, and before I could say anything, the heads quickly disappeared.

'Do you know what John Egan looks like?' I asked Dave.

'No,' was the reply.

'Well mate, you do now, you have just been talking to him.' The two faces had been John (later Sir John) Egan, Jaguar CEO, and Jim Randle, Jaguar's Director of Engineering.

After the all-nighter the team had an early lunch and then crewed for the race after working for 28 hours with no sleep.

This iconic race had the largest entry of 80 cars on the grid, and they were flagged away in two groups. Six hours later and it was a BMW walkover, with the 635is taking the first six places on home ground. Quester/Winkelhock won from Kelleners/Grano, Vanoli/Christian Danner, Bos/Duby/Krucker, Fornage/Haezebrouck/Gartman and Hartge/Nussbaumer/van Ommen.

Walkinshaw had comfortably led the first quarter of the race, but soon after the driver change Nicholson found he had a faulty clutch, the end result being the centre of the drive plate breaking and putting him out of the race. It was an unusual problem, as this component did not normally fail. On the second XJS, a loose oil union following the engine change had to be rectified, and after this cost time in the pits Walkinshaw switched to the remaining car and put in some impressive times to move back through the field. This fine charge came to an end when a chafing wiring loom began to cause the engine to run rough, before catching fire and finally the remaining Jaguar stopped out on the circuit.

Nürburgring winners Quester/Winkelhock.

The TWR Rovers ended up having accident, suspension and engine damage as they tried to prove a point at the Nürburgring and take on the BMWs on their home circuit. There was to be no holding back with the Rover team, they were here to impress and find out where the weak areas were by driving flat out, but be in with a finishing chance by entering three cars. The six hour race took its toll on all three; the Brits were forced to withdraw from this low point in the season and return another day.

Seven hours after the Nürburgring race start, a group of us were in the twin-engined plane that had bought the engine parts to the circuit, returning in the dark and driving rain. I sat alongside the pilot throughout the two-hour flight because I don't travel well, and learned some points about night flying and navigation in zero visibility. On final approach there was some seat-of-the-pants flying when the pilot told me not to look down, but to keep looking straight ahead because that would be where any obstacle would be. I did look down, though – just in time to see some very ornate chimney pots slip under the right wing; we were trying to land on the roof of Blenheim Palace! The dull glow ahead was in fact the Woodstock roundabout lights, not the airport runway lights. I knew exactly where we were, as I passed this point every day on my way to work, and between us we averted a potential problem. With some relief we crossed the airfield perimeter hedge, and the pilot made a last-minute adjustment with the small problem of approaching the main runway from the wrong angle. Some of us went home to bed after nearly two days without sleep.

Between the Österreichring and Nürburgring races a semi-secret test was carried out at Silverstone with Group 44's American IMSA specification XJR-5. Jaguar computer analyst Peter Lings had drafted some goals for this test with the main one being fuel consumption versus lap time. IMSA regulations allowed free fuel specification and no consumption restriction, while the proposed FIA Group C regulations would have limited fuel consumption and octane rating.

Increasingly EngDiv. was being pushed towards using the new Lucas engine management system for this programme, something

I had reservations about. We had also committed to the development of a suitable V12 engine for Group C racing and begun to assemble parts for a prototype 6222cc version to be used in conjunction with Jaguar designed and cast 4 valve heads.

The initial layout drawing for the Lucas engine management system

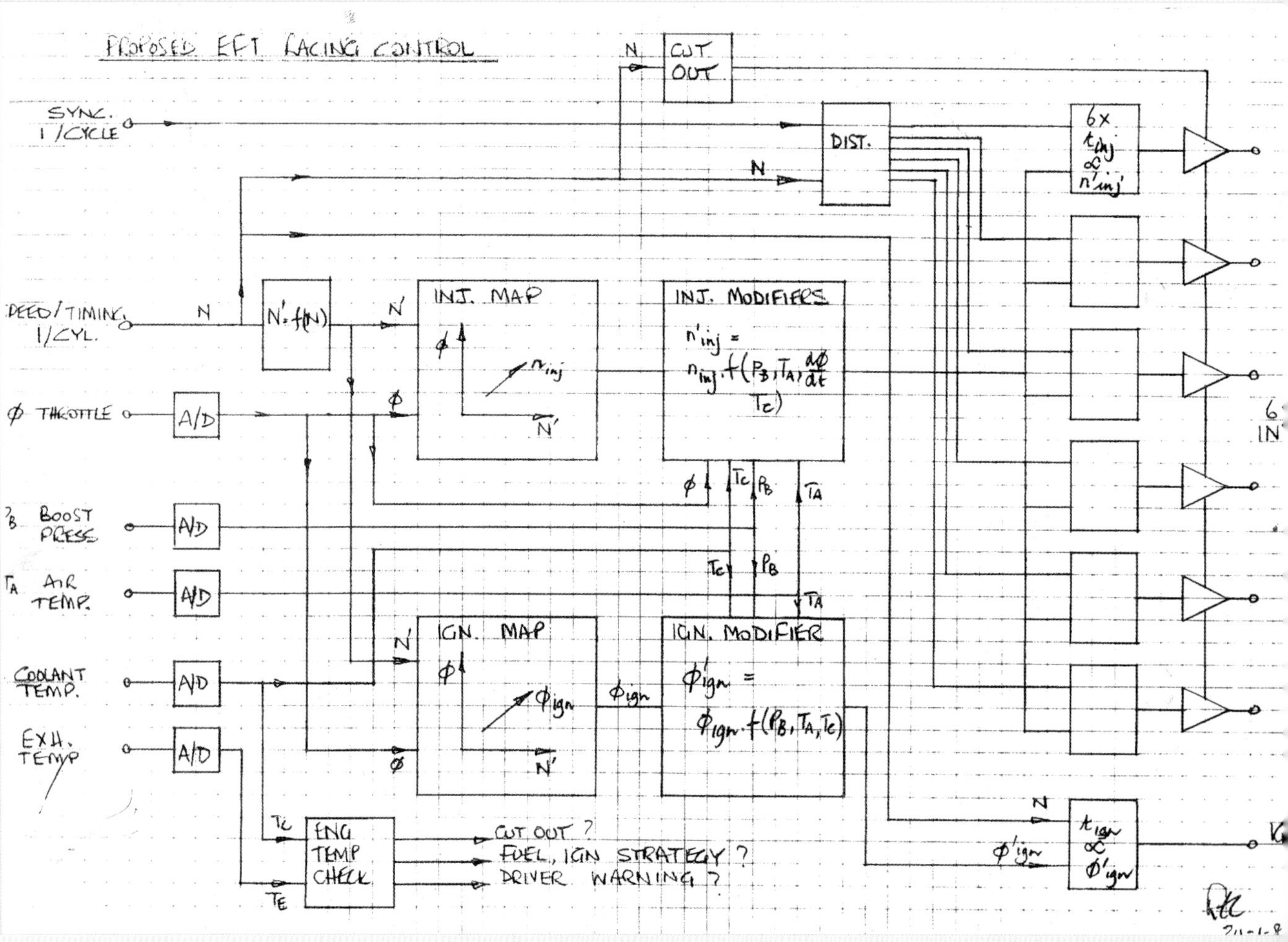

We were now having direct meetings with Lucas on their new engine management system.

MEETING NOTES

Subject: Electronic Engine Management System

Venue: Lucas Micos, 11 April 83

Present:

Mr.J.Randle, BL	Mr.J.Moore, Micos
Mr.T.Walkinshaw, TWR	Mr.S.Goodwin
Mr.A.Scott, TWR	Mr.M.Skinner
Mr.D.Andrews, Lucas	Mr.R.Cattermole

Notes:

1. TWR and Micos to liaise and work directly together in the development and application of the system. BL and Lucas will provide back up as requested.
Progress review meetings will be held at about two monthly intervals. Micos will provide monthly project notes as its normal practice.

2. Micos tabled Draft System Specification for comment by other parties. BL will respond with comments by 22 April. TWR and Micos will discuss system in detail at meeting to be arranged at TWR.

3. TWR expect engine to be running on test bed June/July and hope to run in car August. Micos proposal schedules test bed controller August. Micos to assess possible early use of basic PI controller for ignition work on other available engines at TWR.

4. All Work is subject to confidentiallity agreement between companies involved.
BL and Lucas to discuss policy regarding future marketing of EEMS to other companies.

Copies: Those present plus
Mr.A.Ives, Micos
Mr.G.Johnson, Lucas

XJR-5 VISITS BRITAIN

ON-the-ball racing buffs were able to catch a surprise glimpse of the XJR-5 Jaguar IMSA GTP car which made its first visit to the UK recently.

An even bigger surprise was the sight of Derek Bell lapping Silverstone at high speed with it. (When one realises that Derek Bell's knowledge of the racing car development is second to none, perhaps the surprise isn't so big after all?)

Less than a year has passed since the XJR-5 made its race debut, and this season it has already scored two outright victories in the USA, as well as enough good placings to put Bob Tullius—leader of Group 44 and its G.T.P. race team—firmly into second place in the IMSA championship which still has a long way to run.

Such rapid progress is rare, but it should not make anyone jump to conclusions about the future. Bob Tullius drove the XJR-5 at Silverstone, too, in order to add his comments to those of Derek Bell. Jaguar's technicians, led by Jim Randle, have contributed a great deal to the development engineering that has gone into the XJR-5. Now that it has run on Jaguar's local circuit, its performance can be assessed that much more easily than from data compiled only in the USA, to which the car returned at the beginning of July.

Maurice Rowe, chief photographer of '*Motor*', was on the spot when the XJR-5 made its surprise appearance at Silverstone, where it was assessed by Derek Bell and Bob Tullius.

Britain's top sports car driver Derek Bell was drafted in for this test with the full knowledge of his employers Porsche. Porsche had built a state-of-the-art sports car designated the 956 with its 2.8 litre flat six turbo charged engine, and it was hoped that Bell could give an opinion on the levels of competitiveness. No doubt Porsche had more than a passing interest in the results as well. The test showed that the XJR-5 was probably two seconds a lap to the fuel off a competitive pace.

After this test a 10 month flow chart of milestones in Jaguar's future race programmes was drawn up. These future plans started to remove some focus from the ETCC series but Jaguar were committing to a minimum five year involvement in motor racing and TWR had to pay close attention to these programmes.

Flow Chart for 1983-84

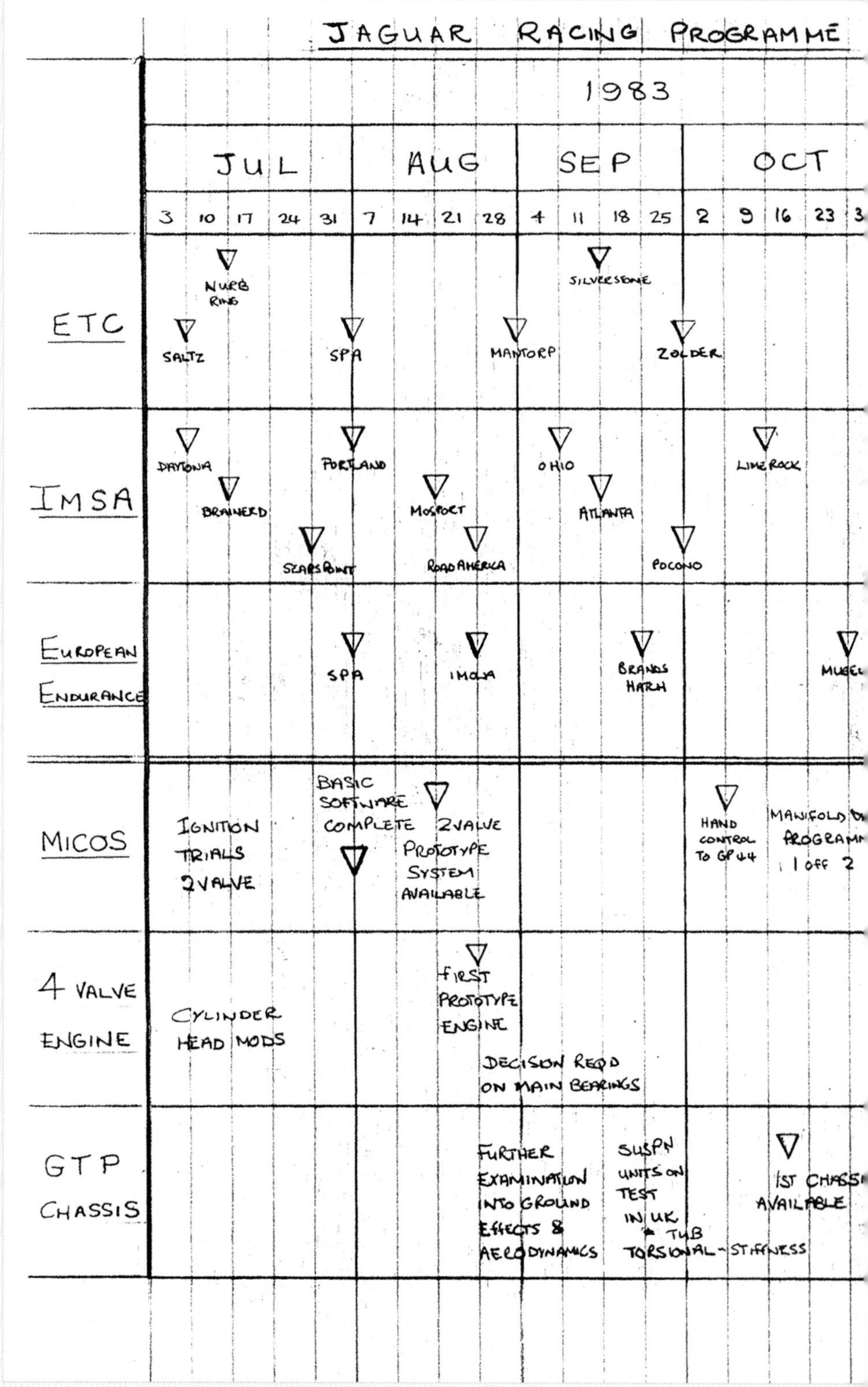

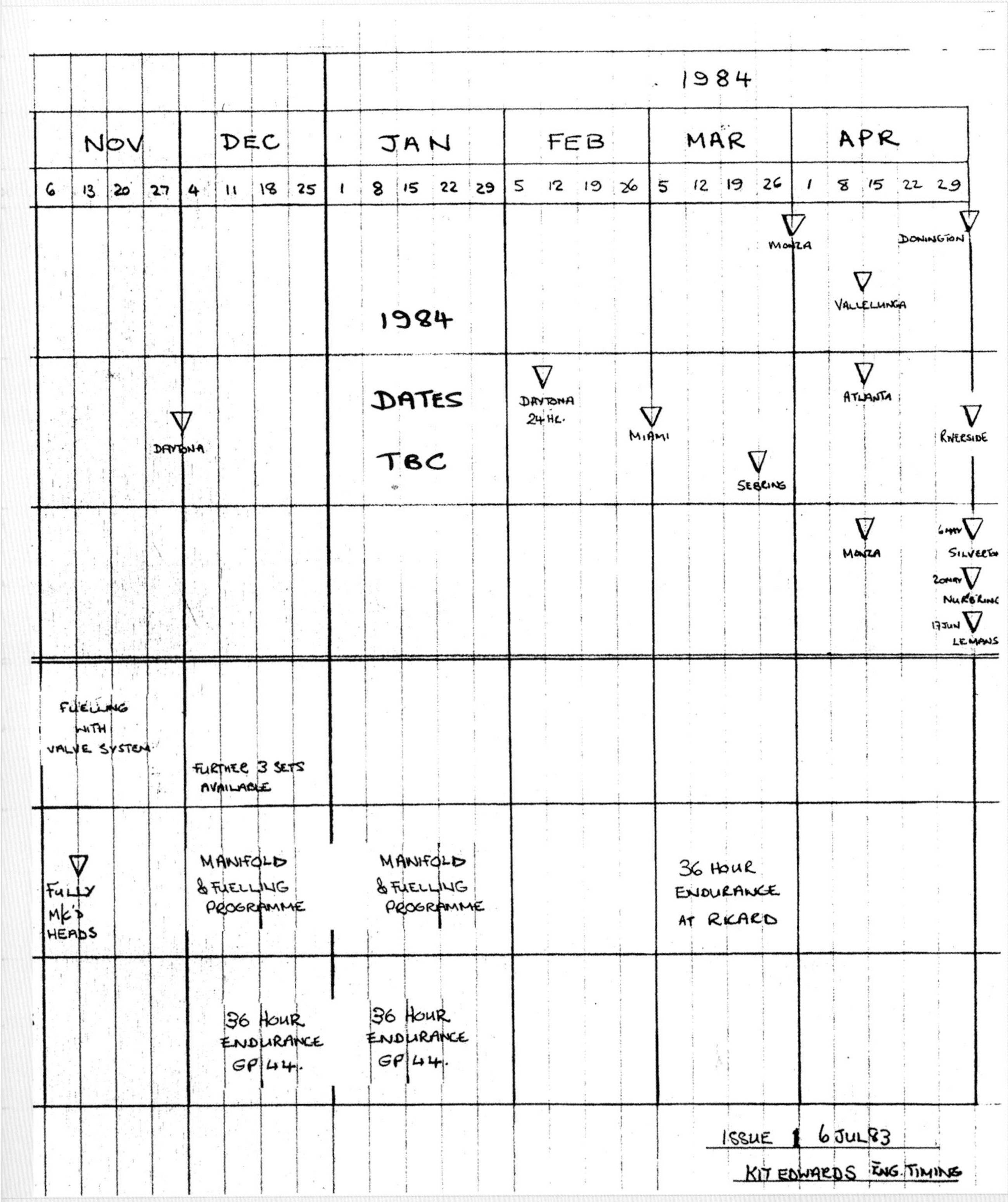
1984
NOV
DEC
JAN
FEB
MAR
APR
6 13 20 27 4 11 18 25 1 8 15 22 29 5 12 19 26 5 12 19 26 1 8 15 22 29
MONZA
DONINGTON
VALLELUNGA
1984
DATES
TBC
DAYTONA
DAYTONA 24 HR.
MIAMI
ATLANTA
RIVERSIDE
SEBRING
MONZA
6 MAY SILVERSTONE
20 MAY NURBURING
17 JUN LEMANS
FUELLING WITH VALVE SYSTEM
FURTHER 3 SETS AVAILABLE
FULLY MK'S HEADS
MANIFOLD & FUELLING PROGRAMME
MANIFOLD & FUELLING PROGRAMME
36 HOUR ENDURANCE AT RICARD
36 HOUR ENDURANCE GP 44.
36 HOUR ENDURANCE GP 44.
ISSUE 1 6 JUL 83
KIT EDWARDS ENG. TIMING

Cigarette advertising in 1983

This was the era of cigarette advertising in many classes of motor racing.

One of the country's to accept a high level of this form of sponsorship was Belgium, where all of these cars would compete in that country's 24 hour race.

Belga and Bastos were Belgian brands, and Gitanes French. Lucky Strike was an American R. J. Reynolds product, and this company would later protest Barclay cigarettes' claim to have the lowest tar content of all. It was then proved that Barclay's filter design was used to circumvent the industry's testing process.

This was an interesting outcome when looked at from a Group A perspective.

Spa 24 Hours – Belgium

The Spa 24 Hours race is the pinnacle of the ETCC series, and is held in such high esteem that many entrants would be just as happy to win that race as the ETCC championship itself. It is the ultimate test of teams and mechanical preparation over 24 hours, in race cars that normally compete for three to six hours of racing. For the winning manufacturer, Spa proves the endurance and reliability of their product, and advertises their engineering superiority to a prospective new car buyer; this appeals to a wider audience than just those with an interest in following the manufacturers' racing programme.

TWR was running two Jaguars plus two Rovers here, and had the welcome return of Win Percy to the team. In conditions that were unusually dry for Spa during both day and evening qualifying periods, Walkinshaw and Percy set the two fastest times and took the front row of the starting grid. This was important as it would gain huge publicity from the hour-long live television coverage of the race start. The Jaguars were followed in qualifying by 10 BMW 635is, a Ford Mustang, and Rovers in thirteenth and fourteenth, followed by another four BMW 635is. One of the problems encountered here was oil temperatures becoming very hot in the warm conditions, unusual for Spa, and additional oil coolers were fitted to help the problem.

Start of the Spa 24 Hours race.

Additional oil coolers were added for the hot conditions.

A total of 57 cars started, but the Jaguars soon lost their lead with the Walkinshaw/Dieudonne car pitting to change a wheel and later a driveshaft, dropping them out of contention. An additional stop dropped the Calderari/Percy/Brundle Jaguar down to sixth place at quarter distance. The Walkinshaw/Dieudonne XJS recovered to thirteenth, but it would retire with differential problems after ten hours.

Just before half distance, the remaining Jaguar had worked its way back to second, the result of some hard driving by Brundle, but soon after Percy was hit while overtaking, and the resulting puncture put his car into the safety barrier. He returned to the pits for repairs, and after replacing a damaged alternator, was in ninth place at the 12-hour mark. With six hours remaining the XJS was in sixth, with Brundle at the wheel, when a connecting rod broke and he was definitely, finally out of the race.

Night pitstop for the XJS at Spa.

The final results were: Heyer/Hahne, Quester/Winkelhock/Rossi (BMW), Allam/Lovett/Soper (TWR Rover), Semoulin/de Dryver/Pirenne (Ford Capri), and Cipriani/Toffoli/Sienna (BMW). Among the many BMW competitors most things had gone wrong that could, including accidents, lost wheels, clutch, gearbox, engine, differential failures and brake problems – but they had safety in numbers.

Third place for the TWR Rover was an excellent result, driven by the experienced and patient UK team of Allam/Lovett/Soper.

Jaguar had never before had a connecting rod break in a V12 engine, so this was thoroughly analysed later at the factory. It turned out that the connecting rod had developed a small folded section called an inclusion during the forging process. It was in a critical area by the head of the big-end bolt, and that had led to its eventual failure, but this was considered by Jaguar to be a one off situation.

Hockenheim – Formula 1 Support Race

Hasty repairs were carried out to both Jaguars and then taken to Germany for the Grand Prix support race at Hockenheim in Germany a week later. This non-championship 15 lap race was won by Enzo Calderari, but Tom Walkinshaw did not finish. It was a great result in front of Germany's largest motor racing audience.

There was now a much needed three week break to repair and rebuild the damaged race cars in time for the next race meeting, the important Silverstone TT on home ground.

ARG Motorsport boss John Davenport as navigator for Tom Walkinshaw had a lucky escape after they hit a tree at high speed in the Bianchi rally.

Silverstone – England

Another connecting rod failed in similar circumstances during testing in the lead-up to the next race at Silverstone before we could have the full report of the failure at Spa. With the season rapidly coming to a close a special manufacturing process would have to be investigated and put in place, but in the meantime it cast a shadow over engine reliability.

The premier saloon car event in the UK was the Silverstone TT and all the Jaguar and Rover senior management as well as the growing number of supporters would be on hand. The Jaguars were split in qualifying by Soper in the TWR Rover, so the real threat to Jaguar looked to be the TWR Rovers on their home track! The Rover team too had been under huge pressure having suffered a virtual write-off of its single rally version of the Rover Vitesse after a huge crash the previous weekend while TW competed in his first Tarmac rally in Belgium. We all of course had to be there as well and it was just another outing that had been added to the programme during the year. That had committed the Rover team to urgent major repairs in time for that car to compete in the Manx rally with Tony Pond at the wheel the weekend following the Silverstone TT.

The race was run in changeable, rainy conditions, and the Jaguars were soon in trouble in this situation that suited the lighter cars. Even after a change of tyres, the times didn't improve and Nicholson eventually crashed out of the race. The second XJS lost two laps while a belt tensioner was repaired, and Walkinshaw (who had not started the race, to give himself the best points option by driving the better-placed car later in the race) then joined Dieudonne and Calderari. On a drenched track he was able to only salvage ninth place, this car/tyre combination not working at Silverstone. A disappointed Jaguar CEO John Egan had missed the expected Manufacturer's prize, and aging Jaguar founder, Sir William Lyons, looked on at what was to be his last view of the racing XJS. Sir William died just over a year later, aged 84, without seeing his cars race again or witnessing the Championship win in 1984.

However, the day was still TWR's as Steve Soper/Rene Metge went on to a fine win in their TWR Rover. The

The winning TWR Rover of Soper/Metge.

so another race was on. We had to win the ETTC with the XJS in 1984, and that would then allow TWR to move into Group C and Le Mans before Group 44 had too much of a lead. Jaguar had spent too much of the little money it had on space frame chassis technology that was soon to be outdated.

As part of presenting an alternative, TWR arranged a visit to the Formula 1 team of Williams Grand Prix Engineering at Didcot, for a demonstration of their wind tunnel. Williams with their design expertise could be an alternative to the Lee Dykstra USA option, introduce Jaguar to Formula 1 technology, and more importantly, provide an indication of ballpark prices. Using expensive Formula 1 technology and prices could then help move Jaguar onto the TWR option that could be a little cheaper, initially.

Mounting pressure for the future engine management system to be the new Lucas Micos prototype surfaced during the later part of 1983. I felt more comfortable with the relationship being developed with ex-Lucas man Bill Gibson, who had been joined by Brian Mason, also ex-Lucas, as a full partner. They had already indicated that the analogue fuel injection solution we were using would become a full digital engine management system, fitting into the standard Jaguar electronic unit. It would also still be able to use the standard wiring loom, with the addition of a small secondary loom. There was plenty to think about towards the end of 1983.

TWR EngDiv. tested the first Lucas Micos 'Synchronous' total engine management system for Lucas and Jaguar at Kidlington. A standard 'HE' 5.3 V12 engine was supplied for testing purposes by Jaguar and fitted with the programmable fuelling and separate real time adjustable test bed ignition system. This system was the brainchild of Roger Cattermole in association with Mike Skinner and others mentioned on page 139, and worked well in this two box prototype form.

Jaguar's support for its various racing programmes during 1983 are listed on pages 158 and 159 to show the level of factory support available in 1983. They include the components for the five special body shells being assembled in late 1983, four of these to be used by TWR in 1984.

With the many different Rover programmes as well as the Jaguar ETCC it had been a year of survival for TWR/EngDiv. The 1984 year would need something special from all concerned to produce a winning Jaguar programme, and fortunately both the chassis department and EngDiv. knew what was needed and had already started on next year's specification well before year's end. TWR had to win the ETCC series in 1984 to then move onto the next Jaguar programme, the prototype sports car Group C.

Decisions taken at Jaguar's monthly meetings had included.

9th August:

USA designer Lee Dykstra is put in charge of the Le Mans programme.

Total Jaguar racing and test programme for 1983-84.

13th September:

Decision made that 4 valve engine will not run at Le Mans in 1984.

Group 44 are looking at making chassis parts as well as tub from carbon fibre. TWR report that these things can be done much cheaper in the UK.

October:

Dunlop's Brian Bennett had been summoned to some of the meetings to progress the tyre weakness that mainly revolved around the wet tyre. We were at a distinct disadvantage in the wet to other teams supplied with European tyres.

TWR's 1983 version of the Group A XJS had failed to win the championship. There had, however, been large steps made in the design by year's end and this would prove beneficial in 1984.

Having initially joined with a commitment to a minimum two year period and now nearly three years into my time with TWR, TW wished to have some security for the future of EngDiv. Over a dinner of jugged hare at The Bear in Woodstock he felt it was a good time for us to buy a house instead of renting. We were in a comfortable situation at the Manor House in Long Hanborough but it made good sense. I accepted his offer of a substantial deposit to help with a house purchase and our home in Dunedin New Zealand was sold; and a longer term commitment made to TWR. There was no employment contract though in the early 80s, all dealings were verbal.

Jaguar too made a new commitment to expanding sales in Germany, taking control of sales for that country via a new company, Jaguar Deutschland GmbH.

Crankshaft torsional vibration test

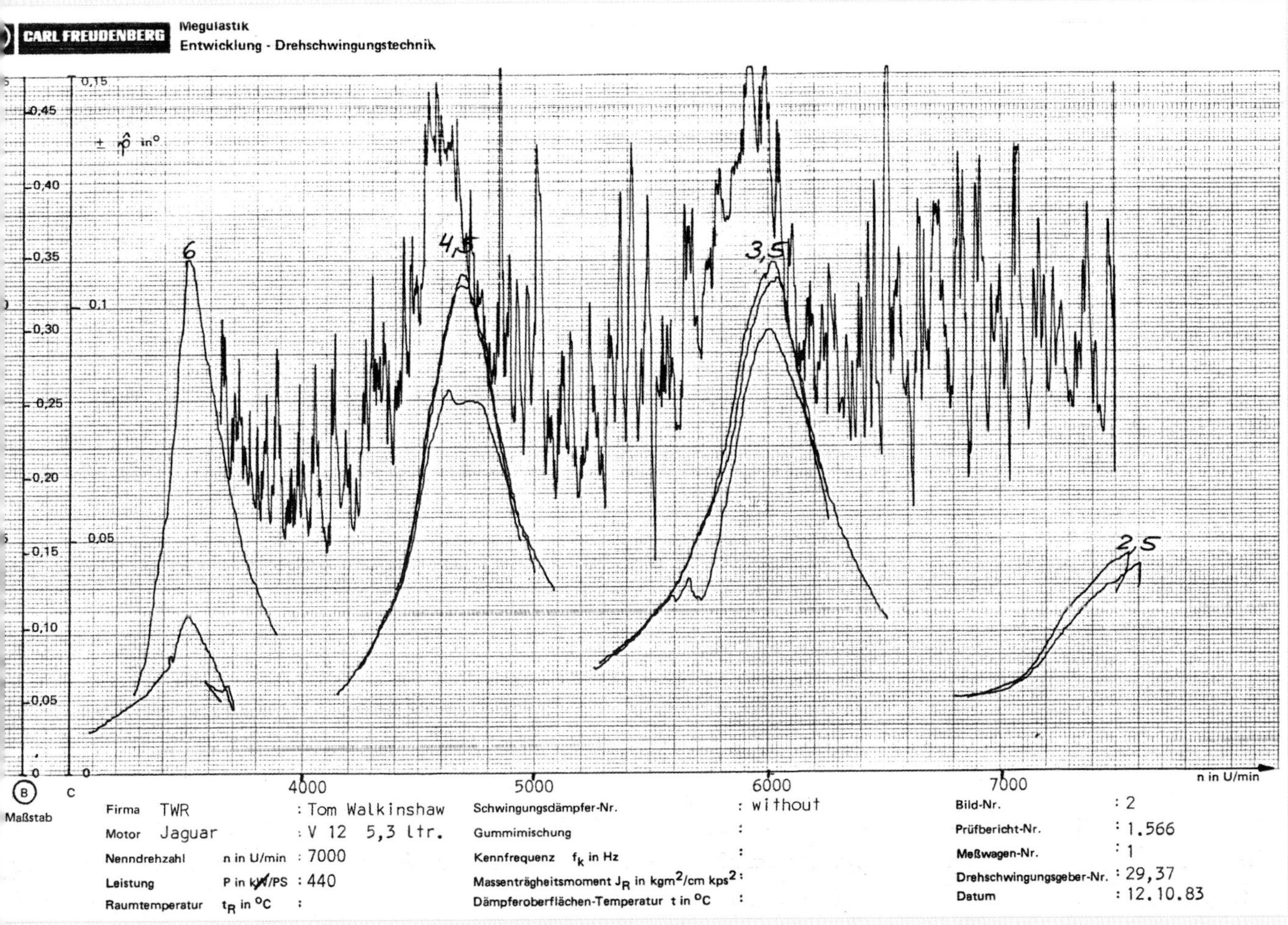

At year's end a crankshaft torsional vibration damper test was carried out by Freudenberg Engineering of Lutterworth to compare the needs if any to be made to the standard torsional vibration damper. The test results are shown above and the standard damper was retained.

Jaguar Factory Parts List 1983

Page 1

Component	Supplier	Order No	Quantity Ordered	Quantity Supplied	Stock	Team Components Supplied to and Remarks
L/W Bonnet L/W Boot L/W Door RH L/W Door LH L/W Frt Wing RH L/W Frt Wing LH L/W Front Lower Wing RH L/W Front Lower Wing LH	Atlas Products Ltd " " " " " " "	EXP/TS/ 505734	10 10 10 10 10 10 10 10	7 5 6 6 6 7 10 10	3 5 4 4 4 3 0 0	2 complete sets supplied to JGR: balance used by TWR. See separate report.
L/W Facia Panel	Bridgtown Industries Willenhall Manufact.	EXP/TS/ 517296 EXP/TS/ 498441	8	5	3	2 supplied to JGR 3 supplied to TWR
L/W Frt Screen L/W Door Glass RH L/W Door Glass LH L/W R/Qtr Glass RH L/W R/Qtr Glass LH L/W Back Light	Triplex Ltd " " " " "	EXP/TS/ 493188	10 10 10 10 10 10	5 7 5 5 5 5	5 3 5 5 5 5	2 sets to JGR 3 sets to TWR TWR had 2 extra RH door glasses due to breakages.
Heated Front Screens	Triplex Ltd	EXP/TS/ 514643	8	6	2	2 off to JGR 4 off to TWR
Rear Sunspension Cradle	Wootton Eng Ltd	EXP/TS/ 519576	10	10	0	4 off to JGR 6 off to TWR The beam comprises thicker S/A mtgs and steel diff mtg spacer plates. Assembled at Radford less end mtg plates.

Page 2

Component	Supplier	Order No	Quantity Ordered	Quantity Supplied	Stock	Team Components Supplied to and Remarks
Oil Coolers 56 ROW AN12 ends	Serck Services	EXP/TS/ 505735	2	2	–	2 supplied to GP44; 6 more to be ordered.
Cam Followers	Farndon Eng.	EXP/TS/ 517269	500	500	0	150 off to FP44; 250 off to Aero Bonner. 100 off to JGR. GP44's were to a Cosworth drwg.
Crankshaft 74 mm x 2.1" pin	Farndon Eng.	EXP/TS/ 514728	4	4	0	4 off supplied to GP44. 1 off at 11 lb weight saving. 3 off at 12 lb weight saving. See separate report re 2½" main brg.
Cylinder Heads 1.80" inlet	Jaguar & Chessman Eng.	EXP/TS/ ?	2 pairs	–	–	1 pair ready for despatch 8/7/83. These heads modified from existing cstgs.
Cylinder Head Gaskets	Cooper Payen Ltd	EXP/TS/ 519626	50 pairs	15	–	13 pairs supplied to TWR; 2 pairs to GP44. See separate report.
Cylinder Liners	Harold Andrews	PROD ORDER	50	–	–	WIP prom 8/7/83 + 040" on crown to enable fit of gas filled rings. GP44

Jaguar Factory Parts List 1983

Page 3

Component	Supplier	Order No	Quantity Ordered	Quantity Supplied	Stock	Team Components Supplied to and Remarks
Frt Cross Beam	Pre Star	PROD ORDER	10	4	6	2 off to JGR; 2 off to TWR. These beams have extra weldir and no spring top hats.
Frt Suspension Hub Assy's	Steering Developments	EXP/TS/ 503106	2	2	0	1 off to JGR; 1 off to TWR. See separate report.
135 L Fuel Cell & Refuelling gear	Premier Fuel Systems	EXP/TS/ 521025	2	–	–	WIP prom 6/7/83. This order for JGR.
Fuel Cell Collector Pots	Safety Devices	EXP/TS/ ?	4	–	–	2 off for JGR; 2 off for TWR. Wtg quote. Required by 19/7/8
Complete frt & rear GPA Suspension	TWR	EXP/TS/ ?	1	–	–	Required for JGR WIP awaiting quote. Prom mid August. See separate report.
XJ-S bodyshell for JGR	RS Panels Safety Devices	EXP/TS/ 517459 EXP/TS/ 519540	1	1	0	Modified shell to JGR instructions and fit Roll cage. Photos are available.
Oil Coolers	Oxford Exhaust	EXP/TS/ 503069	8	8	0	4 off 19 ROW, 4 off 25 ROW } For JGR

Page 4

Component	Supplier	Order No	Quantity Ordered	Quantity Supplied	Stock	Team Components Supplied to and Remarks
Radiator Cores to Broadspeed Spec	Marston	EXP/TS/ 503128	4	4	0	4 off to JGR
High Capacity Oil Pumps V12	Kensun Eng. Metric Eng. Farndon Eng.	EXP/TS/ 517354 EXP/TS/ ? EXP/TS/ 517313	25 25 50 pairs	20 – 20 pairs	5 – 30 pairs	Oil pump cover & body } Gears 12 sets of covers, bodies 7 gears to TWR 8 sets of above to Aero Bonner. Quote required from Metric.
V12 Camshaft with GP44 Profile	Lydmet	EXP/TS/ 521133	20 pairs	–	–	WIP from 14/7/83, then 2 week m/c. 18 pairs to GP44; 2 pairs to TWR. Possible Aero Bonner requirement.
Dural Main Brg Caps	Farndon	EXP/TS/ 519537	4 sets	–	–	2 sets for TWR; 2 sets for GP44. 1 block is being delivered to GKN for brg checks 6/7/83, est weight saving 11 lbs.
Steering Racks	Jack Knight	EXP/TS/ 519597	4	4	0	4 supplied to GP44; different ratios as required: 1 x 7 tooth; 1 x 8 tooth; 2 x 9 tooth
Cylinder Head Studs in EN 24W	Arden Precision	EXP/TS/ 519615	20 sets	–	–	WIP prom 6/7/83. 10 sets for GP44; 10 sets for TWR.
Cylinder Head Nuts in EN 16T	Jaguar	–	20 sets	–	–	Comp. waiting for studs.

The engine bay of the 1984 Group A XJS. Showing back-up ignition pack, large diameter fuel rails and heat shield between inlet manifold and cam covers.

chapter five

Preparing for 1984

With the extensive development period over the winter of 1983-84 at TWR following the disappointing 1983 ETCC season, both TWR's Jaguar and Rover programmes would be considerably enhanced. The goal for 1984 had to be the Championship win in the ETCC, and no excuses. Also, while both TWR ventures needed to appear to be on equal terms, it had now become vital that the Jaguar XJS should fulfil its early promise. Bigger things waited in Jaguar's future, and these needed to be pursued no later than the following year.

The situation of one team being more equal than the other at TWR in 1984 would also need careful handling and TW was the master in these situations, with his stand-alone teams housed in separate buildings on the Kidlington site. The Jaguar team had moved once again into even larger premises just up the road on the Kidlington Estate with JaguarSport now taking these vacated premises.

Towards the end of 1983 EngDiv. had gained the services of TWR's design draftsman Russell Sharp, who had completed most of the drawings for the Rover Group A programme. Now half of his time was re-allocated to the many updates and ideas I wished to introduce to the Group A V12 engine, and this was a big help to me. My updates for the Group A Jaguar had all been written up by October 1983, allowing Russell Sharp to start the outline drawings.

We had also started preliminary work on an experimental 6222cc V12 engine for a future Group C programme and Jaguar's return to Le Mans, with development and testing starting in early 1984. This experimental engine was to be fitted with a

The first experimental 6222cc V12 engine using Jaguar designed 4 valve heads.
Note adjustable exhaust and inlet system for resonance tuning.
First tests produced 670bhp at 7500rpm.

pair of 4-valves per cylinder heads that had been part of a development exercise at the Jaguar Browns Lane Experimental Department. Jaguar's programme had only ever reached the head casting stage and none of the material in stock had been fully machined, let alone tested. One or two people in procurement had some reservations about the quality of the pieces involved but nothing more had been said officially in our meetings. These 4 valve heads needed additional detailed drawings, and Jaguar supplied us their production drawings that could simply be marked up with our changes. While this sounds logical, we still had to contend with Jaguar's in-house intellectual property and security systems that had to be respected, sometimes slowing the procedure.

The drawing support facility for EngDiv. was based in a separate building, so I had to work with a shared system and priorities and this did not always suit my timetable. It was a situation that would continue in the years to come as new programmes and the drawing office expanded, but TW would never approve a dedicated design/drawing area at EngDiv. Chassis and engine design skills do not best come from the same person, but the idea of a central design office was something TW wanted to maintain to best share resources, and perhaps remove some of the considerable independence the Engine division had.

EngDiv. would increase to twelve people during 1984, with Rob Benson and Glenn Barber joining as experienced engine builders. Rob Benson would later transfer to Australia after working for the Aston Martin Group C programme, joining the newly formed TWR Australian racing division in 1990, before a career in NASCAR in the USA. Glenn Barber would much later go on to manage the remaining members of the TWR engine team move to Leithfield some 25 years later. These two were then joined by

Steve Grey, whose name can be found on many of the hand-built Aston Martin V12 production engines, and all employees at this stage would play an important role in the future build up of talent at EngDiv. This had increased the team to seven engine builders in the race division and two more in the JaguarSport road engine department.

With Austin Rover considering a withdrawal from the British Championship series in protest over the way it had been treated in 1983, TWR had to consider a Plan B for the future of ARG Motorsport's Rover programmes as well.

ARG Motorsport had become well financed, with approval to go international Group B rallying with its Metro 6R4. However, problems with the development of its complex rally car delayed that programme and they could not commence their full rally series on time. For TW, this was a budget pocket just waiting to be picked. ARG Motorsport probably needed another high-profile programme to spend its money on and a Plan B in case the ongoing dispute with the Rovers in the BSCC went against them.

This dispute would come to a head early in the year and after an abandoned litigation case in the Old Bailey with TW, a negotiated exit deal would see me left with the task of defending the situation at a tribunal in London. At this enquiry I was cross examined by two prosecution barristers and a panel of experts from the RAC, overseen by Britain's former lead prosecutor at the WWII Nuremberg Nazi trials, Lord Shawcross.

The problem had centred on the steel rockers supplied by ARG Motorsport. The fact that they had been used in 'my' Rover engines meant it was my problem to defend the issue. With this prospect in mind TW offered a deal to run a two-car Rover team in parallel with the XJS Jaguars at

Jeff Allam with the Group A Rover.

each of the ETCC events in 1984. Two rival manufacturers run by the same team had never been seen before.

Mark Snowdon, of Austin Rover, subsequently announced the new Rover Vitesse SD1 racing team to compete in the ETCC for 1984, and that TWR would run the programme. He added that he was confident Rover would be superior to BMW, and win the series: he had, after all, seen the Rover win at the previous year's Silverstone TT. Clearly, somebody had forgotten to remind him that Jaguar's XJS Group A team was also run by TWR – and in the same events.

ARG Motorsport boss, John Davenport, was a skilled presenter and politician to his superiors at British Leyland who were more focused on keeping a nationalised car industry afloat than on the side-issue of motor sport. Davenport also knew he needed a substitute programme until the Metro Group B rally car was competitive. Equally, Jaguar understood it was to be their year after coming so close in both 1982 and 1983. TW would have to be the ringmaster with these two equal teams.

In addition to the Mazda and BMW dealerships at Dorchester and Gerrards Cross, the TWR group now had four factories on the Kidlington Estate with the arrival of the new JaguarSport business.

In January 1984, TWR and Jaguar had formed a joint venture called JaguarSport, initially to enhance the performance and styling

The new line up of JaguarSport XJSs outside the Kidlington factory.

of the XJS. Production V12 XJSs all used the new HE 5.3 litre V12, but needed a complete facelift. Jaguar, however, did not have the funding to completely revamp the ageing design, now in its ninth year of production, but customers wanting something extra could have the kudos of driving a TWR-modified XJS with body and engine enhancements.

The V12 engine programme was added to EngDiv.'s list of things to do, and centred on exhaust systems, cold air intakes and high-performance air filters, plus conversions to 6 litre engine capacity.

JaguarSport was formed as a 50/50 joint venture between Jaguar and TWR, and capitalised at £5m. Jaguar Cars supplied £2.5m in cash, while TWR exchanged its £2.5m liability by EngDiv. supplying four-valves-per-cylinder engine technology to Jaguar for future developments, mainly directed at the V12 engine. The total amortised cost of running the entire engine division at this time was less than £250,000 a year. Jaguar bought back TWR's share of this enlarged business around 10 years later for nearly £17m.

The extra JaguarSport work was to be carried out alongside the race engine preparation, initially with no stand-alone department for this project. The programme was simply handed over to me without much discussion as to where the manpower would come from, or its ultimate aim.

Anthony Moss was recruited to supervise the new section, along with engine builder Mike Collett.

This area set aside at EngDiv. also developed a series of high-performance Rover V8 road versions, one being for Formula 1 boss Frank Williams at Williams GP. Large-capacity V8 engines were developed with TW thinking that these would be needed by British traffic police whose 3.5 litre Range Rover vehicles had become quite slow when filled with the one ton of equipment needed to control motorway closures, making them not quick enough in motorway vehicle pursuits.

EngDiv. was also increasingly seen by Jaguar as an R&D facility and we became involved in the development of a high-performance road engine for the current 4-valve 6-cylinder XJ40 Jaguar, and this had to run in parallel with our race engine business in 1984.

Known as the AJ6 engine, this programme centred on camshaft development for lower and mid-range torque, and used a quill shaft drive system to remove the problematic torsional vibrations

CONTRACT BETWEEN TOM WALKINSHAW RACING LIMITED and JAGUAR CARS LIMITED

RE: JAGUAR HIGH PERFORMANCE ROAD CAR ENGINE

The scope of this contract allows TWR to carry out the following work on behalf of Jaguar Cars:

		COST £
Camshaft, valve train and Drive System Analysis and modifications to include:		
(A) (i) Camshaft and Camshaft Profile Valve Spring system; (ii) Modifications to Drive Assembly to isolate Camshaft and Timing Chain Reversals and reduce noise.		
	AMOUNT PAYABLE	30,000.00
(B) Modifications to Inlet Manifold including Patterns for Castings		
	AMOUNT PAYABLE	15,000.00
It is hereby agreed that Jaguar Cars shall supply TWR with 2 6 cylinder XJ6 complete engines. TWR shall thereafter provide and engineer, one to be modified to include System (A) for Jaguar to evaluate on their Test Bed and the other to include Systems (A) and (B)		
	AMOUNT PAYABLE	10,400.00
	TOTAL CONTRACT VALUE	55,400.00

Notes: (i) Jaguar Cars shall have the right to use all drawings and Pattern work involved in the above project.

(ii) All amounts herein are subject to VAT of invoicing.

AJ6 enhanced engine programme.

from the valve train. Although very successful, it was not carried over into production, perhaps being too expensive to manufacture. The value of this contract alone, which consumed half a man-year of time and after sub contract items, would pay 25% of the total salaries for the engine division. These six cylinder engines were later increased in capacity to four litres by the Jaguar factory to improve performance.

JaguarSport engines would eventually grow, with Nigel Grist and Steve Corbett being added, before this group was handed over to the stand-alone JaguarSport division two years later. After that, JaguarSport engine development continued under the direction of Ian Brown.

Sudden increases in workload were now normal for us, and usually happened after planning for the year was supposedly finalised. TW was also expanding his business interests outside of racing with a Jaguar car dealership added to the existing BMW and Mazda franchises. Sometimes team members were called on to carry out tasks that were outside of motor racing. I remember one occasion that had us standing around Chuck Nicholson's latest Track Marshall tractor parked outside the Kidlington factory, designing a larger fuel tank for it.

We were all aware that TW's ambitions were not confined to just motor racing, and accepted that it was for the greater good.

I continued to be pulled in two directions, with demands from both race teams, effectively supporting two races on the same day at the same tracks in Europe. On the face of it they all had equal priority, but with TW back in his familiar seat as No. 1 XJS driver, some would inevitably be more equal than others.

Managed well though, it could provide some efficiency for TWR. The entire organisation could join together on race and test days, co-operating with travel and logistics. TWR was now by far the largest Group A team competing in Europe. In its third season of European Group A racing, TWR would be fielding five cars in each race – three Jaguars and two Rovers, and representing half of the manufacturers in the top class.

With a final attempt at producing the ultimate Group A specification in mind, in mid-1983 Jaguar had been commissioned by TWR to create four XJS body shells devoid of any trim, unneeded components or bracketry. The emphasis was on reducing weight and removing items not needed for racing. This factory support was needed to ensure the TWR XJS could race comfortably at the minimum weight of 1400kg, an option also used by our opposition teams.

Paul Davis and Kevin Lee inspect the assembly of special XJS body shells at Jaguar's Castle Bromwich facility.

The bonnet, boot lid and door components were first acid dipped to reduce the panel thickness and then assembled at the factory. The front quarter and radiator panels were of lighter gauge material as were both front fenders (wings) *(see Parts List pages 158 and 159)*. The four TWR body shells had a thicker gauge bulkhead panel fitted where the brake

The ride heights were reduced in 1984 to generally lower the whole car, and the front reduced a further 20mm to produce a nose-down attitude. The front valance panel then came in for detailed attention to lower the mounting point for the front rubber spoiler. These changes were aimed at reducing the airflow under the car and in turn lower the high front aerodynamic lift. After these modifications to the front of the car, the gap between the rubber spoiler and the road was reduced by nearly 60mm compared to 1983. The wind tunnel had produced limited results that the team could work with; it really just confirmed our previous conclusions.

Experiments carried out with various limited-slip differential clutch pre-loads using different thickness shims, had settled on 100lb of pre-load. These clutch packs were replaced after two complete race meetings. The standard XJS Jaguar ramp angles were retained. The differential coolers remained as before mounted in the air stream in the rear either side of the fuel tank.

The special 5-speed heavy-duty gearboxes used since mid-1982 and fitted with gears that were 4mm wider and located in a longer sand-cast casing were retained. The team did not maintain these gearboxes; they were again serviced after two complete race meetings by Getrag in Germany. The gearbox cooler was retained in the area below the front windscreen where the original car's heater had been and was fed cooling air down through the grill below the screen. Two cool air intakes were taken from this same area and fed into the driver's compartment to help ventilate the hot interior during the races.

Gearbox mounting and new finned sump design.

As the year progressed and with the additional staff allowing concentration to the smallest detail in the pursuit of weight reduction, another 15kg would be removed to bring the XJS just under the minimum allowable weight. This attention to detail mainly focused on fasteners and brackets, with aluminium and in some cases plastic being used in non-stressed areas. These latest reductions would provide the chance to manipulate weight distribution by adding small amounts of lead ballast to the rear while the race car kept close to the minimum weight limit of 1400kg.

With the production XJS's normal weight distribution biased towards the front [56%], the race car had handled better with a high fuel load in the rear-mounted fuel tank, as this improved the car's front-to-rear weight balance. In 1983, with up to 70 litres of fuel in the tank to produce the best balance during the vital grid position qualifying runs, the race car had then been up to 85kg over the minimum weight allowed when qualifying.

With the new 1984 TWR Jaguars being 35kg lighter, and when filled with just 40 litres of fuel, they would now be 50kg lighter in qualifying as well as 30kg lighter during the races. The front/rear weight balance however stayed the same 56%-44% empty after all the lightweight items were added.

Race strategy would also have to ensure enough fuel was left to be above the minimum weight at the end of the race if no ballast was added and the water tank for the brake cooling was empty.

The fuel tank capacity in race cars competing in Group A would occasionally become an area of contention and the XJS needed to have as much capacity as possible, to allow for any race strategy that might eventuate. Some earlier tanks had been made a few litres over the regulated limit, or the flexible bag tanks could stretch along their seams, so their capacity would become larger than the 120 litres allowed. Any amount over the limit, no matter how small, would mean disqualification. To ensure the fuel capacity was correct whenever this check was carried out, a bladder was placed in the fuel cell and connected to a small rubber hose that ran all the way forward to under the driver's seat. At the end of each race, on the slow-down lap, the driver would blow through the hose inflating the bladder in the fuel cell and then tie it off. This process removed any excess capacity from the fuel tank should it be scrutineered. This capacity check was carried out by filling the tank and then measuring the contents when drained.

After all the effort put into the various components of the TWR racing cars it would then be down to the tyres to transfer all this hard work onto the track. TWR's racing programmes had relied on the UK Fort Dunlop tyre supplier for almost five years. It was the most important partnership that TWR would have with any of its external suppliers.

Detail changes were made to the front suspension geometry to assist the latest tyres, which would run on Dunlop's patented Denloc 'run-flat' rims. The new, larger 17 inch diameter wheels were built

The new 17 inch Speedline Denloc double-bead rim used in 1984.

to special order by Speedline in Italy to accept Dunlop's new double-bead design that would keep the tyres anchored to the rims if punctured. These magnesium alloy wheels were 10 inches wide at the front and 11-11.5 inches at the rear. Tyre sizes were 275/625/17 front and 300/625/17 rear. If the Denloc beads performed as promised, they would allow a reasonably fast return speed to the pits should a puncture occur. Later in the year Win Percy put this to the test at Brno, and they performed exactly as planned.

Dunlop's William [Bill] Mac had been seconded to support TWR's tyre needs at a critical point in the expansion of Tom Walkinshaw's racing business in the later part of 1982. His task was to produce tyre engineering solutions at the race tracks and supervise manufacture and timely supply of tyres suitable for the many different circuits, from the Dunlop factory in Birmingham. He was one of the few who were not overawed by TW, with tyre performance being a constant contentious area at many race meetings. I observed that Bill Mac and TW had a long term mutual professional respect for each other, a situation that only a selected few outside of TWR enjoyed at that level with Tom Walkinshaw.

Bill Mac has supplied some thoughts from that time.

Bill Mac – Dunlop Tyres UK

It was a fast learning curve that started with TWR at Mugello in 1982. I was brought into the Race Division at Dunlop some nine months earlier to commission and programme a new radial tyre machine. My expertise was writing programme instructions for the operator that defines all the machine settings to enable a tyre to be produced to the required specifications.

The division manager, Alec Meskell asked me if I could go to Mugello explaining that all the regular service engineers were unavailable.

(Later I found out that their feed back was not good enough. One of them famously saying that 'Parc Ferme was in a muddy field' denying him information on the state of the tyres at the end of the race!)

I think the fact that I was not a motorsport enthusiast was a big positive as I was never over awed with what was going on around me and just stayed focused on the tyres required and how best to persuade the team to get the best out of them.

The big decision at Mugello in 1982 was whether to use the hard tyres, or the very hard tyres advised by the previous engineer. The weekend however was all about the engine which had a problem in practice and if I remember it did about a stint and a half before failing [valve spring broken]. I learnt a lot and got on well with Tom, Paul Davis and the rest of the team. I left thinking about how we could improve things for the following season.

Over the 1983 and '84 season with Tom's support we pushed the boundaries by using BTCC sprint compounds in ETCC 500K races. I now understood the tyres' capabilities with regard to the delicate balance of suspension camber angles, ambient temperature and tyre compound as well as tread thickness. Marry that with different driver styles in the same car and the need for the tyres to last for a stint and not too many laps more, and the stage was set for some very anxious but exciting races. I used to stand in the pit lane at the start of the race willing a cloud to cover the sun just for the first 8 laps or so to increase the odds of the tyre not over heating when they were at their thickest tread depths.

Soon my manager Alec Meskell was questioning my requirement for the sprint compound 832 in my Race Requirement list. It became increasingly difficult to persuade him that it was the way to go. Tom however started to order the tyres direct from Dunlop acting as a different customer and this worked for a while until Dunlop race tyre distribution was then handled by TWR/DART, enabling greater supplies directly to the new company set up with Tom Walkinshaw. Come 1984 Alec Meskell agreed to go along with my compound requirements making life somewhat easier!

However Alec was determined to try and break the good relationship between myself and TWR, refusing to pay me to attend a round of the UK BTCC and resulting in me going in my own time and cost. This was not lost on Tom who did everything he could to persuade Alec that continuity was a good thing.

Tom was a very different driver to most others, he seemed to brake much harder and enjoyed the higher front slip angle with the more forgiving capabilities of crossply tyres.

His preferred set up for clockwise direction circuits was a hard left front, medium right front, medium or soft left rear with a soft right rear. This combination allowed him to push hard in to a corner leaning on the brakes, then feather out with good traction off the corner.

I remember Macau. In 1984, Tom started on the 832 sprint compound and we were both worried, but knew we had to take the risk. Half way through the race there was a big accident which formed the field up opposite the pits. This was good for two reasons, the tyres got to cool down and I was able to check them and tell Tom that he could drive without concern for the rest of the race. He won the race, as I am sure everyone else involved with the task, felt we had as well.

The Dunlop story would not be complete without mentioning race tyre fitter John Clifford, his bright smile throughout his huge track side tasks in fitting tyres for the three car team with hand cutting of slicks during races as weather patterns changed was met without complaint.

He is still alive at 83-years-old [2011], and until recently president of his local golf club. He was one of the best tyre pattern cutters and an enthusiastic tyre fitter and all-rounder. I had to have one of the tyre debeaders made extra long so he could break the beads of the rims as he measured up at just less than 5ft tall.

Alec Meskell – Manager Motorsport Division
Roger Jenkins – Engineer
John Clifford – Tyre Fitter / Tread Cutter
John Ward – Tyre Fitter / Tread Cutter
Graham Shuter – Fitter
Tony Radford – Fitter
Brian Bennett – Chief Designer

There was considerable support for William to be employed at TWR when the Dunlop contract was terminated later in 1990. William wished to continue his interests in aerodynamics in the Jaguar Group C programme, not tyres, but team manager Roger Silman did not want this and so he did not join TWR.

Engine

The V12 Jaguar engines were considerably updated and modified for 1984. Developments that had been planned since the later part of the 1983 race season had proceeded through the winter break to become stage three of the V12 Group A engine evolution. Major redesigns for 1984 included the oil sump and cooling system, exhaust and inlet manifold systems and camshaft profile. This would increase the power output and address some oil lubrication problems that had reappeared owing to the increasing race car grip levels.

Another continuing problem had been the high temperature of the sump oil, this heat soak being transferred to the water in the block which needed to be kept below 95°C during racing, and was another reason for the re-design. The sump oil could reach 140°C, transmitting its heat through the conductive aluminium side-skirted block to the water jacket surrounding the cylinders. This heat soak transfer would also produce a rapid increase in block water temperature after the engine had been turned off, quickly leading to boiling and loss of some water coolant through the radiator pressure cap. This situation had to be avoided in races such as the Spa 24 Hours in case a minor repair meant the car had to be stopped for more than a few minutes. If sent back onto the race track in this condition the car would soon begin to overheat because the hot oil had caused the water in the block to boil.

The greater expansion of an aluminium engine block, compared with traditional cast iron, increases engine bearing clearances when hot. With seven main bearings being fed under pressure with hot oil, a considerable amount of oil was constantly being bled from this area, and finding its way back to the sump at the rear of the long engine was difficult under racing conditions. Also, oil pumped to the top of the engine to lubricate the two camshafts had a long return path, and this oil was subject to the changing G forces during racing, producing a sporadic return to the sump.

New wet sump design

My completely new design to fix these problems was drawn up by Russell Sharp and we then cast an all-new higher-capacity finned aluminium sump, which was further strengthened to withstand the inevitable occasional trip over kerbs and chicanes. The sump was made up of two separate pieces, the main upper section, with cast baffle plates, and a separate oil sump pan bolted to it, allowing easy removal and cleaning. This new sump pan had external cooling fins that extended below the front suspension sub-frame line, protruding into the air stream and acted as a very effective oil cooler in itself.

Rather than cool the hot sump oil via two coolers before it entered the engine as we had done in the

Finned and strengthened sump pan.

Excess oil produced when the pressure relief valve opened at 55psi passed through an internal drilling and across to the secondary sump recirculation circuit on its pump intake side. This additional relieved oil was re-circulated and positively returned back to the oil pickup pipe in the sump pan.

Later when it was found that the finned aluminium sump was very efficiently cooling its oil, and to simplify the pipe work, the recirculation pump delivered the oil directly back to the sump, and the pressure side ran through the two oil coolers.

Alternative option used by other competitors

Another idea in use at the time by other teams to eliminate oil surge was the spring-loaded oil pressure recovery system, which worked via a sealed diaphragm inside a tube with a nominal 1 litre capacity. When the engine was started and coming up to pressure, oil was pumped against the diaphragm, which then compressed it against a large-diameter spring and the diaphragm moved back in the tube until the spring and oil pressures were equal. This created an oil reserve under spring pressure that would remain there until surge caused a momentary pressure drop, allowing this oil reserve to be pushed back into the oil gallery by the compressed spring, and maintain the correct oil pressure. Later, on various parts of the circuit where the pump had greater capacity than the engine needed, the additional oil would be pushed back into the oil recovery system. An idea such as this could have also been used on the Broadspeed XJC V12 engines, but never was.

This spring-loaded 'oil pressure recovery system' could be homologated and a poor-quality photograph was attached to the XJS homologation application for the new 1984 oil circulation system. The new 'oil pressure recovery system' was duly accepted. While we could have used such a system as previously mentioned to produce acceptable oil pressure control, it did not have the added needed benefits of cooling the sump oil.

Frontal view of the sump showing the steering rack and exhaust system.

The new design allowed the V12 engine to be raced with better separation between the crankshaft assembly and the sump oil, while maintaining the original oil capacity because the new sump pan was set a little lower in the airstream than previously. Keeping these two items apart reduced oil churning from the crankshaft, lowering both temperature and oil aeration.

The line over the chicanes. Staying inside the main kerbed area. Monza 1984, the last race for the 'rear oil cooler'.

The result of all this was a lowering of the oil temperature in the sump pan by 30°C and more importantly, having cooler oil enabled the oil pump to increase its pumping capacity and maintain oil pressure. Previously the hot sump oil had been cooled after the oil pressure pump, reducing its efficiency.

The XJS's oil temperatures were stabilised at an oil gallery temperature of 100°C, and 110°C-115°C sump temperature, and with the water operating at 90°C this now eliminated the radiator venting problem when the engine was switched off. There was now just a 20°C difference between sump oil and water jacket temperature, as opposed to the previous 50°C.

Additional extensions were later welded to the bottom of the cast baffle plates, with a corresponding spacer plate fitted between the sump and the top housing, and this was a fine-tune of the system. This lowered the sump pan a little more into the airstream to aid cooling and added another litre of oil capacity if needed. The heavily finned sump pan was designed to cope with the possibility of being driven over the concrete chicanes, and the sump-lowering kit, already manufactured as a backup, was only added early in the season after it became apparent that none of the drivers were attacking chicanes and kerbs, the boss included!

I had dreamed this system up some time before but we did not have the time or ability to make it happen in 1983. Thanks to Russell Sharp's help with the casting drawings for the moulds and overseeing the machining process, it was now a reality. The bottom end of the V12 was over engineered in terms of its bearing capacity, and the bearings could probably last a full season if the lubrication system could be made bulletproof.

This would enable us to leave the crankshaft in place on most rebuilds and just check the bearings without having to replace them or repair scuffed crankshafts. It reduced the rebuild turnaround

From the left: Charlie Bamber, Mathew Norris and Peter Grace.

times by 20%. Leaving a rear main bearing rope seal alone if it was working perfectly was another good reason behind this. Along with the slip-fit liners that could simply be replaced in the event of a piston-related problem, it now became an easier unit to service, especially with the inevitable late nights involved. Shorter turnaround times with good predictability on engine delivery to the Jaguar race team was the aim, avoiding any delays and downtime because of the need for outside machining; we still had no in-house machining facilities.

Importantly, this time saving could be channelled into the application of Jaguar's 4-valve heads on the prototype V12 Group C racing engine as this programme developed through 1984.

The 1984 XJS V12 engine programme, supporting three cars and including the Spa 24 Hours race, was supported by just one full time builder, Charlie Bamber, assisted by Mathew Norris, with some other help when time constraints needed fast turnarounds. This programme was completed using a total of only six engines for the three race cars. The final cost of this engine programme for 1984 was two thirds of TW's budget forecasts.

Following our problems at the Nürburgring in 1983 with the standard off-the-shelf crankshafts from Jaguar, we had sourced standard semi-finished EN 16 steel crankshafts, with oil drillings deleted on the production line, in late 1983. With Tom Seal's help we out-sourced completion of the final machining, to higher-than-production tolerances, including cross-drilling the big-end-bearing journal oil feeds, and finishing off with longer heat treatment to increase surface hardness. This work was done at Farndon Engineering in Coventry.

The crankshaft modifications, with signed-off inspection to tolerances of 0.0003 inch in taper and ovality at Farndon Engineering helped speed up engine builds with good security of parts as well as giving longer bearing life under racing conditions.

Detail work on the piston crown shape saw the compression ratio raised to 12.25:1, thought to be the limit for the 98 RON/88 MON low-leaded general European fuel, these pistons being supplied by Cosworth.

Working with Mahle in Germany, we designed new aluminium liners with Nikasil-plated cylinder bores to reduce internal friction on the rings and pistons. They saved nearly six kilograms in engine weight, but failed to produce any improvement in power output. They were expensive and if even slightly damaged had to be thrown away or returned to Germany for time-consuming replating, which did not suit our uninterrupted rebuild process that was necessary to meet the demands of the racing schedule.

Mathew Norris assists Charlie Bamber with the Group A V12 engines.

I was also determined to ensure no repeat of the connecting rod failure of Spa in 1983 and a second one soon after in testing at Silverstone. Using parts that have been mass produced to a price carries the risk of one-off problems in high-stress racing situations, and investigation had shown the weak point in this case was an internal inclusion, probably caused by forging the metal when too cold. This would fold the metal over in the stamping process of the forming die, trapping impurities beneath the surface, instead of forming very hot material perfectly in the die as it was stamped.

Production specification for the factory connecting rod material was minimum EN 16, and Jaguar was asked to supply EngDiv. with forgings made from the blanks of the more suitable EN 24 material. The blanks were forged in the original Jaguar dies, and deemed legal because the Jaguar forging drawings only stated EN 16 as a minimum – meaning there was no upper limit on material quality.

The special batch of connecting rods received individual treatment throughout the many production stages, with each operator station being made aware of TWR/EngDiv. and Jaguar's special needs. An important area, the big-end bearing bore, was left machined undersize allowing this critical area to be finished to a higher tolerance at Farndon Engineering.

The original big-end nut-and-bolt arrangement was discarded, and instead the connecting rod was threaded to take special bolts

The new connecting rod (right) had the stud and nut arrangement removed and was fitted with Umbrakco bolts.
This allowed a smooth radius in an area that was previously weakened with spot face machining.

supplied by Umbrakco in Coventry. Production connecting rod weights varied considerably over the years with replacement dies and new suppliers, and we carefully measured hundreds of connecting rods at the factory to choose a set for re-homologation. Obviously we only chose the lightest connecting rods, and once again this new weight would be put through the homologation process as an 'erratum', or previous error. This selection process resulted in a con-rod weight reduction of 85 grams. It was also an example of the good support and working relationship that had developed between EngDiv. and Jaguar Experimental procurement.

A full digital engine management system was now about to become available through Jaguar from Lucas's experimental division. EngDiv. had begun carrying out testing for Lucas competitions on this system, known as the Lucas Micos full digital engine management system, early in the year. I was now coming under pressure from Jaguar to pick the system up and push its development forward faster than Group 44 in the USA, who had been given the task of track testing the system. Change was something Group 44 founder Bob Tullius, a former military man, approached with some caution, and they appeared unwilling to move too far into uncharted waters. These development programmes would inevitably have meant some added costs, and possibly reliability problems, hence Group 44's reluctance to push its use in racing.

The two separate units for fuel injection and ignition had over time been increased in size as additional heat-soak materials were added. They were, however, never designed to be retrofitted into Group A regulations, but for Jaguar's Le Mans/Group C programmes of the future. There was nothing wrong with the hardware; I just didn't want to be involved in a committee situation, as well as probable restrictive contracts on any use in future projects (including production vehicles). With access to a small flexible workforce at Zytek we were also gaining valuable knowledge that could be applied to all future TWR engine projects.

Because this Lucas system was to become a synchronous (timed single-pulse injection) system, I suggested to Jim Randle this was outside of the Group A engine management rules, and since it also did not fit inside the original electronic control unit, he agreed not to push for it in the Group A programme.

This was a situation that produced the first possibility of problems between EngDiv. and Jaguar, and needed to be handled carefully. Jaguar mentioned its traditional trio of trade supporters – Dunlop, Castrol and Lucas – but my feelings were that 'two out of three ain't bad' as we would move towards a relationship with Castrol later.

The Zytek analogue fuel injection system had been further modified for 1984 and was now a much improved system. The original system had the one full load map being adjusted up or down as the throttle was actuated and that had produced lean mixtures at lower and rich mixtures at higher rpms on part throttles.

We now had a system with eight unique throttle opening maps adjusted via the throttle potentiometer using a 0.5 [off] to 5 volt, [full throttle] signal. This compensated for the varying fuel needs to a much more accurate level, improving driveability and throttle response. This analogue system would then reach the end of its development cycle by mid-1984, after just two years in production. Bill Gibson and Brian Mason of Zytek were working on a new digital microprocessor-controlled system but it would not be available until the later part of 1984.

Some small deviations in the external shape were introduced to accomplish the desired internal routing, and new inlet manifolds to homologated inlet and outlet dimensions were cast and machined.

Most would consider this a step too far with rule interpretation, but I had spent a lot of time walking the pit lane during race weekends to observe teams' reactions to my looking at their engines. Whenever I went near our competitors' cars in the pit lane, the mechanics seemed to hastily close the bonnets, leading me to conclude there was something they did not want me to see on their engines. My suspicion later centred on the 635i's inlet manifold, and this was eventually confirmed when I received some information on special parts (*see chapter 8, page 253*).

Exhaust system

The cast exhaust manifolds used up to this time had groups of three cylinders exiting straight into a common chamber, normal practice for production engines.

An exhaust system fabricated from mild steel tube had been built to replicate the tight confines of the cast exhaust manifold and then tested to determine any differences produced by small variations in these lengths, as well as secondary and tail pipe lengths which were free under the regulations.

With the compromises known, we manufactured new cast replica exhaust manifolds, four to each engine. These were a completely new and complex design, to keep the three cylinders separate internally before joining at the manifold exit point. The new exhaust manifolds were to the exact sizes stipulated by the homologated entry and exit dimensions, and machined to retain the standard heat shields. These exhaust manifolds were responsible for the very distinctive V12 engine sound heard from the start of the

Triangulated exit of the three internally separated exhaust systems.

1984 racing season. It made the cars even more appealing to the public and media, as well as producing worthwhile increases in power outputs.

Camshaft

The Cosworth camshaft EA3/4 profiles used in 1983 were in for an update too, and Chris Walters Engineering, who had been used for the Rover V8 programme, was our preferred specialist. Up to this point, all the 1984 improvements to inlet and exhaust manifolds using the Cosworth EA4 profile had increased the 1983 power outputs by 35bhp, to 465bhp at 7200rpm, with no increase in maximum torque.

With Tom Seal's help we re-cast Jaguar cam carriers, adding additional material to the base and discarding the spacer plate previously used, raising the camshaft centreline to accommodate the proposed very large camshaft lobe base circle diameter of 0.740 inch to be used with the new profile. A rapid opening and closing camshaft profile designed by Chris Walters was helped by fitting larger-diameter 1.450 inch cam followers to accommodate the definition of the design profile and these were sanctioned under reclamation of the cam carrier bores with the need then for oversize followers. This high speed profile operated at the very high maximum valve acceleration rate of 0.010in per degree of camshaft rotation. The profile produced a long dwell period around maximum lift where the valves were held open within 1mm (0.040in) of maximum lift for 110° of crankshaft movement. The entire valve package was redesigned using Cosworth Formula 1 DFV valve springs and was rated to a maximum speed of 8000rpm. Total opening duration was 296°.

When assembled at 100° lobe centres, the profile created a valve lift of 0.222in at TDC, with a maximum operating lift of 0.413in, 100° ATDC. It would hold the inlet valve open within 1mm (0.040in) of its maximum lift, from 50° ATDC to 25° BBDC. It was the ultimate in low-lift rapid acceleration camshaft design, with a totally unique shape because of the valve lift restrictions, and was designated WJ1. These new camshafts were manufactured from EN 40 steel.

We finally selected the best compromise power curve by tuning the lengths of the two secondary pipes running from the two exhaust manifolds on each bank of the V12 engine before

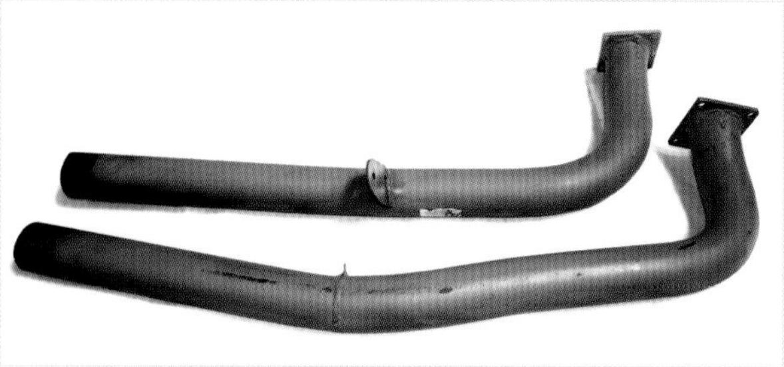

Secondary exhaust pipes

merging into one exhaust pipe for six cylinders. This proved to be a very effective second tuning source. Tests were carried out on secondary lengths of 32 to 44 inches, using two and a quarter inch outside diameter tubing. While the 32 inch length improved the power band below 5500rpm, the best overall compromise was at 36 inches.

The two secondary exhaust pipes from each side of the engine were sand-packed and carefully hand-bent reproductions made by exhaust specialist Len Hartley in the TWR Jaguar workshops. They merged with single tail pipes running either side of the gearbox, which were then ducted out of the left side of the car to keep the unsilenced exhaust noise away from the drivers.

Exhaust testing showed the tailpipe lengths had little effect on power output, and they were really just a collector for the tuned-length secondary pipes. Final results produced a tail pipe diameter of three and a half inches for each bank of the engine, an increase over the 1983 specification which had used three-inch-diameter tailpipes. The new exhaust system benefited from these tailored secondary lengths, to produce an improvement throughout the normal racing rev range of 4500 to 7500rpm.

Ignition

Maximum ignition timings of 32° at maximum torque were now being used, with manipulated distributor advance curves delaying maximum total advance of 35° until 6500rpm with the 12.25:1 compression ratio. The 4.2 XK injectors continued to be used but with the increased power out puts the original circular fuel rail did not have enough capacity while running the group firing of six injectors at a time. It was replaced with two separate larger diameter fuel pipes and these shared a common fuel delivery entry and then exiting through a common 3 bar pressure regulator.

The 1984 V12 engine specification finally produced an additional 55bhp over 1983, a 13% improvement, and could rev out freely to 8000rpm. Nearly 35% of the total power increase came from fitting Chris Walters' WJ1 camshaft and the fine tuning of the secondary exhaust pipe lengths when used with the new inlet and exhaust manifold systems.

chapter six

1984 ETCC challenge

Monza – Italy

Martin Brundle, fresh from testing his new Tyrrell Formula 1 race car, was drafted into the TWR team for the first race of 1984, at Monza, Italy. TW had helped Martin out financially when he had first driven in Formula 3 with the Eddie Jordan team in 1983 towards year's end when his budget had run out. Brundle had just missed taking the Formula 3 crown in that close battle with the future world champion, Ayrton Senna, and he appreciated that help. For TW it was a very good investment too, with Martin wistfully mentioning he had come on board for Monza without discussing a driving fee.

Brundle had to quickly re-adjust to saloon car driving techniques and get used to slowing 1500 plus kg of race car instead of the 600kg of his Formula 1 car. In an early practice incident he did not even try to turn into the entry of the new Monza slowing chicane at the end of the main straight, but drove straight over it. He offered this explanation: 'At 160mph at the end of the straight, I decided to be careful and brake a little early on my first fast lap, but I was already 50 metres too late!' The new cast sump with in-line external fins had passed its oil-surge tests earlier, and Martin had just subjected it to its second test on the Monza concrete kerbing, and now everyone was happy with the 1984 oil-system specification.

The first dry qualifying session announced the arrival of TWR's two manufacturers' teams of three Jaguars and two Rovers for 1984 in great style, with its five cars in the top five places! The Jaguar driver pairing for Monza was Walkinshaw/Heyer, Percy/Nicholson and Calderari/Brundle. TW's new co-driver, Hans

The two new full time drivers for 1984, Win Percy and Hans Heyer. Note the unbranded overalls in this early official Jaguar photo. David Sears would later also join the team.

Heyer, had large business interests in asphalt and the surfacing of autobahns, with 150 of his own trucks and another 150 on subcontract. Later in the year he would take one look at the resurfaced Spa circuit and say that the wrong system and material had been used, and he was right; it had to be re-laid soon after.

Hans Heyer was well established in his own right in European motor racing circles, and a top racing driver. But he was not only bought into the TWR Jaguar team for his talents, his nationality was also a significant factor, with Jaguar out to increase its sales in Germany. Heyer, who lived in Bavaria, could establish a following in his home country through his exploits with the Jaguar team. This would increase the team's coverage in the German motoring press, helping to counter the patriotic German media who were inclined to favour BMW, just as the British media favoured the Jaguar.

The pole time set by Walkinshaw was three and half seconds faster than the sixth-placed car, the Volvo of Lindström/Simons. Percy/Nicholson were second and Calderari/Brundle fourth. The

Early laps in the wet conditions at the first race in Monza.

two TWR Rovers in third and fifth were having a good series debut, as they turned their focus from the British to the European series.

The race though would not be as straightforward as the qualifying pace had indicated. Changing rain conditions throughout the 87-lap, 500km race prevented the Jaguars from extending their lead, and a delayed wheel change during the first pit stop actually saw the Walkinshaw/Heyer XJS lose contact with the leaders for some time. After a final stop for fuel and tyres on a drying track Walkinshaw managed to claw back a 52-second deficit to regain the leading position just 12 laps from home.

Crowded pit lane at Monza.

He was followed by the BMW 635is of Kelleners/Brancatelli, Sourd/Dorchy, Grano/Muller, and Cudini/Vermeersch, followed by the Lindström/Simons Volvo in sixth.

Martin Brundle had set the fastest lap time, but a stone punctured the radiator and eliminated his car. The third XJS retired with overheating after throwing a fan belt. The XJSs were still running the analogue fuel injection system with its delayed rev limiter, and a missed gear shift was probably the initial cause of the problem as well as a much more freely revving engine. We had completely revised the belt drive system and this was a disappointment.

After that first race, TW decided to do away with the contentious 'rear oil cooler' to silence the mutterings of competitors who protested the 'coolers' were outside the rules.

Location of the rear suspension and elimination of any flexing and wheel geometry change had removed the last elements of instability in this area. TWR simply had a much better car this time and with its improved balance and handling the loss of the cooler was not missed. It had been TW's baby and he was reluctant to admit it was not needed; it had been a clever idea rather than a good one. It also did not hurt to give the opposition teams something to focus on at the rear of the car either, instead of them trying to

Jean-Louis Schlesser.

decide how the front spoiler was now much closer to the ground than before. Because of the winter development from EngDiv. too, the XJS was now very fast in a straight line and able to pass any BMW 635i with ease.

Vallelunga – Italy

The tight Vallelunga circuit near Rome in Italy saw Walkinshaw and Percy qualify in the first two places, but the third XJS, driven by Schlesser this time with Calderari, was caught out in the changing track conditions which produced rainfall at the wrong time for them.

Versatile French TWR Rover driver Jean-Louis Schlesser had joined the Jaguar team to partner Calderari, since Brundle was not available. Schlesser quickly settled into the task. They had joined many of the other top cars experiencing the same problem with the changing conditions, and were back in seventeenth.

The 157 lap race had 32 starters, and with rain falling, the slower-qualifying but faster cars soon began working their way to the front, which made for a more exciting race. Intermediate tyres fitted to the leading Walkinshaw/Heyer Jaguar were not good enough in the changing conditions though, and that XJS began drifting back and losing places. Percy remained in contention, and with a changeover to Schlesser at his first stop, they continued in the top three.

About mid-race, things changed for the entire TWR crew. A jammed fuel hose in the BMW pit next to us sprayed fuel over the pit wall and road, and the panicking team next door quickly waved their car out. Unfortunately the BMW's side exhaust exit sparked a fire, and in the chaos a Goodyear technician checking rear tyre temperatures slipped over and was covered from head to foot with burning fuel. He was completely engulfed in flames, but that car's pit crew had fled and it was up to the TWR team

Cramped pit lane at Vallelunga.

(who had all been trained in fire-fighting by the Kidlington Fire Brigade) to put out the fire. Meanwhile the flames had spread to BMW's tyre stack, and as the 200-litre refuelling tank became surrounded by flames, things started to look very serious.

We also attended to the burning Goodyear tyre technician, who had all the clothes on the lower half of his body burned off. There was no support from track officials, and in a further farce the ambulance driver, who had stationed his vehicle at the best observation point for his own benefit, promptly dug his vehicle up to the axles in the soft earth as he tried to rapidly reverse out.

Enzo Calderari lost three laps before he could return to the circuit.

Meanwhile the race continued, and just when it looked as if the Percy XJS would win, it suffered alternator problems, lost the battery voltage and was out. The Walkinshaw/Heyer car suffered with overheating front brakes from damaged brake cooling ducting, but finally took third place. Cudini/Snobeck (Juma-Bastos BMW) were the winners, with the strong pairing of Stuck/Quester second in the Schnitzer entry. The remaining places were also taken by BMWs, with the third Jaguar of Calderari/Nicholson coming eighth.

BELOW
From left:
Andy Smart, Blue Dorwood, Ron Elkins, Allan Scott [author], Paul Davis, Keith Partridge [side on],Russell Sharp [back], Kevin Lee, Gary Davis and Dave Roberts [back].

Donington – UK

The third round of the series was back to Donington, the TWR team once again enjoying the luxury of a home race with less travelling for the team as well as enjoying the parochial British support. The supporters were rewarded in Saturday's qualifying with a Jaguar 1-2-3,

The Jaguars lead the pack in the opening lap of the race.

along with TWR Rovers in fourth and sixth. UK drivers Weaver and Woodman had the fastest BMW, in fifth position. With British drivers in the top six places this result would produce a large turnout of fans through the Donington gates for the next day's race.

However, once again the Jaguars would have to cope with one more pit stop than the BMWs, as well as conserve their brakes on this tighter circuit, while in the cooler British conditions the Michelin tyres of the one-stopping cars would be at their best. As the race settled into its first phase, the Jaguars ran at the head of the field until a bad tyre vibration saw Percy pit for fuel and four new tyres, allowing Weaver's BMW into second place, followed closely by Schlesser in the third Jaguar and Granberg in his increasingly competitive Volvo.

TWR provided the JaguarSport pace car for the Donington round.

At half-distance the Jaguars were first, third and fifth, split by Weaver's BMW and Steve Soper's TWR Rover. A broken wheel on the Walkinshaw/Heyer XJS saw the end of that car's run, because when the wheel was replaced it was seen that the air ducting to the front brakes was torn off, and time lost in repairing this would remove the car from contention. Brake overheating once again dropped the number one XJS out of contention, and it eventually finished in ninth position.

Second place for the all-British Weaver/Woodman team.

Percy and Nicholson went on to regain the lead and win the race, with Weaver/Woodman second, Stuck/Quester third and Grano/Muller fourth, followed by the Schlesser/Calderari Jaguar in fifth. The poor result still left the Walkinshaw/Heyer pairing at the top of the points table, however.

Enna Pergusa – Sicily

There is something unique about each venue, and the next one at Enna Pergusa had a few things, including its squat toilets in the rural Sicilian setting, and the availability only of bottom-of-the-range rental cars, Fiat 600s. On the way to the track in these hire cars we discovered we could get another three to five kph of top speed when driven in fourth gear rather than fifth and this was just all part of normal travel, seeing who could get to the circuit first.

The fast open Enna circuit with its central lake.

The Sicilian circuit surrounds the shores of a bug and snake -infested lake, and the teams arrived just in time for the annual hatching and flight of small insects that soon coated the windscreens of the cars, reducing visibility almost to zero after 30 minutes of practice. Various hastily improvised fly deflectors began to appear during practice and qualifying. They partially succeeded in leaving a large enough gap on the windscreen for the drivers to see straight through.

By now, a few of us were fully paid up members of the 'Sod's Law Club', established after our second places in the 1982 and 1983 Championships. With the competitiveness of the XJS, TWR's three cars and drivers were all capable of winning, producing the potential problem that without team orders, the points would be shared among the three cars. This meant the Jaguar team would not enjoy a decisive lead over the closest consistently finishing BMW. I felt we needed team orders to ensure the championship was won before the last race, because if we didn't have it in the bag by then something was sure to go wrong during the last race, and we would lose the whole thing again. There were ways however that TW could move the situation in his favour without being too obvious.

The new cars ran without power steering as in 1982 and

Start line with fly deflector fitted to the bonnet of the XJS.

1983, even though most drivers preferred to have it. They also had a hard brake pedal because of the DS11 brake pads and the larger brake master cylinders to help reduce a long brake pedal. These placed more physical demands on the driver's arms and legs, but it also suited TW's physique. However there was one change made as a compromise; the original XJS brake pedal had a 3:1 ratio when used with its original power booster system, but now the pivot points were changed to 4.5:1, providing increased leverage on the master cylinders and helping reduce some of the driver effort needed. TW's seat position was set exactly as he wanted it, and then welded in position for 'safety', meaning that his taller co-drivers would have to live with this seating position. Generally it was TW who chose the chassis setup, and small adjustments to front roll bar stiffness would be the only changes drivers would make to suit their own driving style to improve balance and 'turn in'. TW therefore had few team orders, believing the drivers had to have a chance of winning or they wouldn't be at their consistent best. The entire team had to be in a top competitive mode and supplied with cars able to win races, especially if TW's Jaguar had problems.

After Walkinshaw was delayed during this race, Enzo Calderari took the lead and was on for the win, but his chances of winning the ETCC were slim-to-nothing, with that position being pencilled in for the boss. Instructions were given to Enzo to slow and let the Walkinshaw/Heyer XJS catch up and overtake. Enzo's father was managing his son's pit signals, and while he duly held up the board with 'slow' on it, he held his free hand out over the Armco barrier, out of sight, and waved his son on – just as any of us would have done in the same position, I suppose. But in any case, Enzo must have been struggling to see

Ron Elkins, nearest camera, looks over the pit stop at Enna Pergusa. Note the TWR fire marshall Loubor Shiller (in helmet) with suitable fire protection, but to his left, partially obscured, fuel vent man Keith Partridge in shorts and shirt.

much out of the fly-splattered windscreen, and no doubt this was the reason he carried on at speed to take the finishing flag.

The Championship was on the back burner again, and the 'Sod's Law Club' was gaining more recruits, with the Calderari/Brundle XJS finishing ahead of Walkinshaw/Heyer and Percy/Nicholson. It was a 1-2-3 for TWR, but not quite in the correct order. The other finishing positions at the end of the 202 lap Enna Pergusa race were the BMWs of Kelleners/Brancatelli and Stuck/Quester a lap down. The Volvo of Lindström/Olofsson was a further lap down in sixth.

The first 1-2-3 of the 1984 season at Enna Pergusa.

This result was discussed by telephone when Jaguar's Engineering Director Jim Randle phoned me on Monday morning, and he conveyed his congratulations on a result he had been waiting for since Monza.

Brno – Czechoslovakia

The trip to Brno in Czechoslovakia was always different. Communism and the Russian influence were still in control in the early 1980s, making the atmosphere a little different from other venues. The locals too were divided into those who accepted the Communist system and those who didn't. All were employed, one way or another, and had money, mainly because there was nothing to spend it on in the few shops that existed. Additionally, the local constabulary was out to fleece us 'wealthy westerners' on arrival each year. We deliberately drove at much slower speeds than normal, for obvious reasons.

On this occasion two carloads of our team were stopped on the motorway and an official claimed all our passports and demanded rather a lot of cash to return them. He had greatly exaggerated our speeds, and after watching our British timing crew make many failed attempts to retrieve the situation, I decided to try my 'colonial' diplomatic skills as a solution to the impasse.

The discussion had been conducted largely in sign language, supplemented by claims and counterclaims as to speed written in the dust on the boot lid of one of our rental cars. I tactfully pointed out that the Czech traffic officer's allegation were slightly exaggerated, and rubbed out his claimed speed to replace it with a lower one. I also emphasised that it was not necessary for him to take such a hard line; we would certainly not repeat the offence. However, something was obviously lost in the translation. The look on his face said he was far from happy with my suggestions, and in an instant he wiped the boot lid clean with a single sweep of his forearm, which was as thick as my thigh. He left just enough dust to scribble an even higher number than before. I took this to mean negotiations were at an end, because there was no dust left in which to print my counter proposal; this was probably just as well for us all.

A mumbled apology, new Jaguar team shirt with my name on the pocket (far too small for him), two ball-point pens, ten postcards in last year's XJS white with green flash colours, and we were handed our passports back and were on our way. The obsolete XJS cards were not as good a deal as I had at first thought: I later travelled from one side of Brno to the other in a taxi and needed just three of them to pay the fare.

Exchange rates were three to four times better from the man behind the tree in the dark at the rear of our Brno hotel than what had been offered at the airport. The team could cash their Sterling allowances and live very well during their stay, even after allowing for the fact that local prices were specially inflated just for the race weekend. The Brno race was a weekend when the boys were able to let off some steam after all their hard work to date, and the local girls were always very friendly too.

Brno was a long and fast street circuit with a substantial

Entering the Bosonohy village after the 165mph downhill straight. This daunting area was where Tom Walkinshaw showed his talents.

racing history. It hosted an international event each year, and those trapped behind the Iron Curtain had this one chance to look at what the westerners were up to. It was expected that as many as 120,000 people would come for the 57-lap 386-mile race.

Brno's street circuit was lined with Armco barriers that left little room for error from the downhill pit straight, which then swept into a blind left-hand fifth-gear corner through the village, with its houses built to the edge of the footpath. The row of double-storey houses on either side less than two metres from the road's edge making this high speed corner blind, led to the use of turn-in and apex points referenced by different colours of front doors on two separate houses. There was no room for error, and it was a circuit where TW enjoyed many successes. Added to that, the open space and fast corners suited the XJS well.

A confident Tom Walkinshaw – in the background I am speaking with FIA scrutineer Marcel Servais.

Qualifying produced another first, second and third grid line-up for the Jaguars. Calderari was joined by a young David Sears, their car being second fastest, ahead of Percy/Nicholson. Sears was the son of well-known former racing driver and at the time Silverstone circuit chairman, Jack Sears. Sears senior later promoted TW as his successor at Silverstone, but this appointment would later cause conflict with the circuit owners, the British Racing Drivers' Club.

The Stuck/Quester BMW qualified next, followed by the Volvos, which were starting to show their competitiveness on this fast circuit; they were fifth and sixth.

The race quickly settled into a pattern with the Jaguars first, second and third until debris from another competitor's crash resulted in a punctured tyre for Percy, which dropped him to seventeenth. This incident did, however, prove the worth of the new Dunlop Denloc tyre and wheel technology when Win managed to take the long drive back to the pits at reasonable speed on the flat tyre, without it coming apart and damaging brake cooling ducting.

By two-thirds distance the Jaguars were again in the first three places until Calderari's gear lever broke. He managed to continue

1-2-3 for TWR Jaguar at Brno. Newcomer David Sears is on the right.

and maintain his position, with the following BMW of Quester having tyre trouble and unable to catch him. The final result was another great clean sweep for the three TWR Jaguars and a win for Walkinshaw/Heyer. The XJS trio was followed by the Volvo of Granberg/Elgh, and the BMWs of Kelleners/Brancatelli and Stuck/Quester. Allam/Duez, in the TWR Rover, finished seventh.

Leaving Czechoslovakia's airport could have its problems too. Brno is famous for hand-cut lead crystal, and the customs agents wanting 'private export tax', with baggage checks and duty demands. The TWR crew were also subjected to a very private frisking in the special customs booth, with curtain pulled, complete with scrotum squeeze in my case. Team members all emerged slightly subdued and quickly got onto the plane for home.

Österreichring – Austria

Moving onto the two Austrian races, starting with the Österreichring at Zeltweg on June 17th, the Jaguars again qualified in the top three positions. Walkinshaw/Heyer went on to win from Percy/Nicholson, but the third XJS ran low on oil and the engine expired before it could be topped up during its scheduled stop. This was the Jaguar team's only internal race engine failure of 1984.

The remaining places were taken by the Grano/Muller BMW third, from the Volvo of Lindström/Olofsson, then the BMWs of Stuck/Quester and Kelleners/Brancatelli. The TWR Rover of Allam/Duez was seventh.

Jaguar pace car making good use of marketing opportunities at Salzburgring.

Salzburgring – Austria

The Salzburgring 141-lap, 372-mile race, two weeks later, saw Walkinshaw on pole again, followed by Percy and Calderari, with the first BMW of Berger/Ravaglia nearly two seconds behind the pole time in fourth. Next were

1950s Jaguar racing manager Lofty England stands to the left of the two winning crews at Salzburgring.

Granberg in the Volvo, then Kelleners, and Stuck in their 635is, each separated by less than a tenth of a second.

Percy/Nicholson won this time, followed by Calderari/Sears, a lap ahead of Kelleners/Brancatelli in third. An inadvertent serious test for the V12 engine was carried out by Hans Heyer when the leading Jaguar suffered a gearbox problem and jammed in 4th gear. Heyer repeatedly took the engine over maximum rpm by slipping the clutch out of the slow corners, but eventually the heat became too much for the flywheel, which came loose and finally broke free from the crankshaft. A large section of gearbox bell-housing broke when the clutch and flywheel disintegrated. I was quite impressed though that it was the flywheel that had broken, bearing in mind that the valve train was designed for a maximum of 8000rpm.

When discussing the one-two result with Jim Randle over the phone the next morning, his question when I did not report another 1-2-3 was: 'Oh dear, what happened? What happened?' We had only come first and second, that's what had happened.

Nürburgring – Germany

If that was a disappointment, then Nürburgring in Germany was a disaster. This track was never kind to TWR and its Dunlop tyres, no matter what car we raced there.

The Jaguars pulled off their now-normal qualifying spot in the usual order at the front, with a one-second margin over the rest of the field. The Rover team displayed their best form to date in fourth, sixth and ninth, meaning there were six British TWR cars in the first nine starting places, at a German circuit. This was a very powerful statement from the TWR teams, all the more so for being made at the home circuit of most of the top competitors.

Roger Silman seen here talking with Martin Brundle.

When the cars had lined up on the grid before the race, Paul Davis took some delight in reminding newcomer Roger Silman that this was a winning team after the latter instinctively turned to walk towards the rear of the grid. He had joined us from the Toleman F1 team, which had struggled for a couple of years towards the rear of the Grand Prix field. It was a new experience for Roger to walk to the front of the grid.

The leading Jaguar, with Heyer starting to allow Walkinshaw a chance to change to another XJS if necessary to pursue his best points outcome within the team, soon returned to the pits with belt problems, caused by stones from an off-course excursion. Repairs to the belts delayed the XJS, while the Volvos were looking good in second and third early on, but then were delayed later when problems set in.

Calderari's XJS stopped dead with a total electrical failure, and Heyer then handed over to Sears, who later retired as a result of the previous overheating. Nicholson kept out of trouble down in eighth place, but Walkinshaw put in a great effort to lift the remaining XJS to fifth – a critical placing for his championship position. A bad day was tempered with the best result to date for the TWR Rover team, with Allam/Duez coming in second, but taking championship points from Walkinshaw. The BMWs of Gartner/Cudini and Hollinger/Giroix were third and fourth.

Spa – Belgium

Two weeks later and it was the long distance Spa 24 Hours endurance classic in Belgium. TWR decided to run just two Jaguars (004 and 006) for this premier event, while the Rover team entered a full complement of three cars.

Nothing was left to chance with the Jaguars' preparation which now included the recently homologated larger 14 inch front discs. The water injection to cool the front brakes would typically use 25 litres of water in a three and a half hour race and it was hoped that these larger brakes driven with some care would last the full distance and the water injection would not be needed. Servicing it with additional water during the 24 hours would take too much time.

The two XJSs duly took up their usual leading positions on the front row and they were followed by five BMWs and then by the first Rover in seventh. The TWR Jaguar driver line up was Walkinshaw and Heyer joined by Percy, and Calderari with Sears and Belgian Teddy Pilette in a one-off drive. That addition continued TW's philosophy of always having a Belgian driver in the team for the Spa race. Not only could they speak fluently with any officials who might not have the best English, but in marginal decisions during the event race officials would see a local man being penalised, not just a British team.

Critically for Jaguar Cars UK, the company was about to be finally floated on the London Stock Exchange. Jaguar would no longer be part of a British nationalised group, but return to its own original identity under public ownership. All were hoping for a win from Spa as a good omen for Jaguar's stock market debut.

After some negotiations, Jaguar's funding relationship with TWR changed at this time also, with the enhanced but mainly pay-per-result basis changing to a fee-based contract with a performance bonus. This was no doubt a situation TW needed because of the envisaged future move with Jaguar to Group C Prototype Sports Cars. Inevitably this would also mean a big learning curve once again in a new series, with wins not expected in the beginning when up against the dominant 956 Porsches, so a contracted retainer produced more stability for the company.

JAGUAR TO GO INDEPENDENT SOON:

"WE'RE ALL DELIGHTED!" SAY JOHN EGAN AND JAGUAR FOLK AROUND THE WORLD . . .

"It's the culmination of four years' hard work by everyone connected with the company". So said John Egan, Jaguar's chief executive, following the official announcement to the House of Commons by Industry Secretary Norman Tebbitt, on 22nd May, that BL could dispose of its shares in Jaguar, by means of a public sale offer. "We're delighted that plans also exist specifically to involve employees in the flotation", Egan added. "It will give everyone who wants it a stake in our future".

Sir William Lyons (82), founder of Jaguar Cars Ltd. and the man who approved Jaguar's original, brilliant, Le Mans participation in the 1950s, is very much an Egan man. "I'm very, very pleased", he said.

Sir William Lyons, FRSA, RDI, DTech., founder of Jaguar.

Early laps in the rain.

Spa, in the mountainous Ardennes region, is notorious for its changeable weather, and 1984 was no different. Most of the cars on the starting grid wore various wet tyre combinations when the race commenced at 5 pm on Saturday, two hours later than the traditional starting time. The heavier Jaguars, with their Dunlop tyres, suffered again in the wet during the first part of the race, leaving the Rovers and BMWs to battle it out at the front.

Antwerp diamond merchant Eddy Joosen put on a good show in his TWR Gitanes sponsored Rover. He maintained this TWR built car with his own small team while competing in the Belgian national series, however it was run with the TWR Rovers from time to time in the European series. His presence ensured a strong TV coverage for the local driver as he battled with Hans Stuck in the 635i BMW in the early part of the race.

It was some time before a drying line started to favour the big Jaguars and by the six-hour mark Walkinshaw's XJS had moved back into the leading position. All was not OK though in the TWR Jaguar camp, with the second XJS already out of the race, having been caught out with the return of a rain shower. When passing a slower car Calderari had left the track and damaged the front suspension. Although he managed to return to the pits, the damage was too severe to repair, and the Jaguars were down to just the one car after only three hours.

Congested area behind the pits during the race.

More showers came and went during the night, but with

Evening pitstop. Note rear window tape to shield following cars' headlights.

the inevitable accidents and pace-car laps, luck turned against Walkinshaw. His return to the track after a stop for wet tyres had coincided with the two pace cars coming out and the pit lane being closed, until the first pace car and all cars behind it passed the pit exit. He lost almost a complete lap.

The team was not happy at all as the Cudini/Snobeck/Tassin BMW took this to their advantage along with the lead. More problems persisted with the pace car control system as Walkinshaw was mistakenly identified as the leader later on that night and blocked behind one of the two pace cars, allowing the leading BMW to gain almost another lap as it sped around the circuit to attach itself to the rear of the train of cars behind the second pace car! Because of these two problems, the clearly faster Jaguar was in third place at half distance, when by rights it should have been a lap in the lead, and the weather conditions were deteriorating once again.

The battered Joosen/Pond TWR Rover was on for second place, but a miscalculation in the fuel saw it stranded out on the track.

Early dawn saw a dense fog descend on the track, causing a two-hour full-caution period with the field having to maintain their positions until forced to pit for fuel. By 8 am racing resumed and over the next three hours a great drive

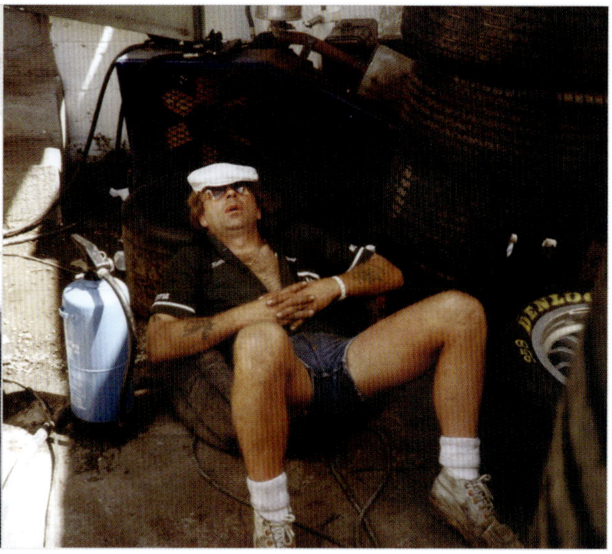

ABOVE
Tom Walkinshaw did not like anyone to go to sleep during the 24 hour race, but in the early morning it was time for a quick nap.
Left: Graeme Hall and author Allan Scott.
Right: Truckie and re-fueler Keith Partridge in his fire suit.

by all three crew members in the XJS moved them back into the lead. With six hours of the race remaining, the TWR Jaguar team became a little more relaxed as the XJS circulated three laps ahead of the BMWs, all in the next five places. The first TWR Rover was back in seventh, after the Joosen/Pond Rover, which had been leading the race at one time, had a miscalculation and ran out of fuel; it was on for a certain 2nd placing.

The Jaguar steadily built its lead over the remaining hours of the race as attacks from the opposition faded, but just when it looked as if nothing could stop him, Walkinshaw was back into the pits just minutes from the end to change a punctured tyre.

A late pit stop for the winning Jaguar at Spa.

Thankfully the strong driving combination of Walkinshaw, Heyer and Percy was very difficult to match, and their hard consistent driving bought the XJS home in first place, three laps ahead of the next car, despite all the vicissitudes of laps lost and a puncture.

As with all these 24 hour races, emotions are high after the huge effort involved, and mixed with total exhaustion the release is a totally unique experience, even for the most hardened team member. It was a great relief for all to tick some very important boxes with this result. Jaguar winning its first 24 hour race for over 25 years and in the same week it debuted on the stock market. Vital championship points had been won, the first three places in the ETCC were now being held by TWR Jaguar drivers. The longer term potential for Jaguar to continue to use race tracks to promote its product to the world had been reinforced, and this would allow TWR also to continue to improve its growing profile and future prospects when advancing into Group C.

The Walkinshaw/Heyer/Percy XJS covered a total of 453 laps, ahead of Cudini/Snobeck/Tassin on 450 laps, Stuck/Quester/Weaver 449 laps, and Thibault/Gartner/Guitteny 449 laps. Spice/Martin/Martin, (Toyota) 443 laps, Grano/Muller/Giacomelli 442 laps, Giroix/Krucker/Hollinger 440 laps and Lovett/Duez/Schlesser (Rover) 437 laps.

Unexpectedly, on Monday afternoon a telex was received at TWR headquarters in Kidlington, congratulating us and Jaguar on a fine win. It came from the BMW factory and was the only official communication TWR ever had from BMW over this period of racing in Group A from 1982 till 1986. It was a very satisfying moment for TW, a former BMW driver.

The teams used the midsummer break to good effect to completely rebuild both the Jaguar and Rover teams' cars. They were worn out after Spa, and were completely rebuilt to continue on in the quest for the Championship over the last three races.

The championship positions now were: Walkinshaw 165; Heyer 155; Percy 126 and Kelleners/Brancatelli on 109.

Increasingly now Group A engine development was about engine management systems to gain ultimate power outputs and improved fuel economy utilising very high compression ratios – one of the main engine areas that was free. The collaboration with Zytek would be a major contributor to TWR's increasing superiority in this area. By the end of 1984 EngDiv. had an in house digitally programmed engine management system to control both fuel and ignition timing, vital when using standard fuel and this would be used in the Rover programme in 1985.

We had tested a first prototype of the digital engine management system on the XJS after Spa. TW drove the car, gave me an indifferent report, suggesting it needed more work and I was disappointed with his view. He had seen the small benefits from my dynamometer tests at Kidlington, but this was how TW worked sometimes, occasionally delivering information to suit the way he wanted to proceed based on his feelings about the situation and his considerable experience. He usually had an overall team view as well when reporting on track situations. Sometimes when we were on top of things at the circuit his engine would mysteriously become 'a bit fluffy', just to give us something to do.

The new digital complete engine management system had a programmable rpm limiter built into the system, which reacted instantaneously, and also gave EngDiv. better control of rpm used at the race track, and elimated lost belts with missed gears.

Despite the advances, TW wished to continue with the original system for the last few races of 1984. He knew the XJS could still win using the improved detail to the older analogue fuel injection system and distributor-controlled ignition. There was also a change in thought patterns as we achieved superiority and had a very fast race car. We had to get it right this year to move on to Jaguar's Group C sports car programme.

He was understandably cautious about new fangled electronics, believing that 'if it isn't broke, don't fix it', with a small input from 'Sod's Law' always in the back his mind too. At EngDiv. we had spent considerable time producing a consistently rising and falling engine torque curve and the improved analogue fuel control programming system was able to react well to these demands.

Silverstone – England

Silverstone was a circuit the team knew well from the many tests carried out there, and its open spaces suited the Jaguars too. We had high hopes of pleasing the large UK crowd that turned out for the Tourist Trophy meeting – the oldest contested race in Britain then and now, having been first staged in 1904 on the Isle of Man.

Again the Jaguars produced the usual qualifying positions, including an amazing run on qualifying tyres by Walkinshaw to result in a time three seconds faster than anybody else. His 1:34.68 gave the locals something to think about. It was just the second full year of Group A racing in the UK, and an amazing effort considering the perceived wisdom three years earlier that the cars

The TT was first raced on the Isle of Man to avoid the 20mph speed limit in Britain at the time.

'Gordon Bennet' was the name given to this 1904 'trial'.

The TWR starting grid for the Silverstone TT.

would be a little slower than the 1982 UK Group 1.5 specification which had qualifying times in the 1:39s. It was a TWR lock out of the first five places on the Silverstone grid with the slowest of its six cars present in eighth place.

From the start the race ran perfectly to plan, with all three cars circulating at the front, followed by the three TWR Rovers, which spent some time fending off various Volvos and BMWs. Six TWR race cars were leading the combined efforts from Europe on home soil, much as they had done earlier in the year on German soil.

With all the opposition just stopping once, the real position started to emerge at around one-third distance after all three Jaguars had pitted for fuel and fresh tyres. It only took 20-25 seconds to change drivers, tyres and to refuel via the gravity-fed tank in the pits. With the in-and-out time in the pit lane, this all added up to just over a minute lost to the one-stopping cars.

The first two XJSs quickly regained their earlier positions, while David Sears brought his car back into third place by mid-distance. This meant all three Jaguars had only one remaining stop, so they were now effectively on the same one-stop schedule as their competitors.

Lap 60, however, changed all that, as torrential rain forced two of the XJSs to come in for tyre changes. This presented a problem because the normal pit setup could only handle one car at a time in each of the Jaguar and Rover teams; there were not enough team personnel or pit space to service all cars at once. The desperate crews were overloaded when all six cars needed to come in from the flooded track for tyres at the same time. What had looked like a walkover for TWR/Jaguar now turned into a disaster.

After having to complete another lap to ease the pit congestion, both the Nicholson Jaguar and the Duez Rover slid off the circuit and into the catch-fence, where they remained with other cars completing the same manoeuvres, their slick tyres uncontrollable in the rain.

Meanwhile, with the pace car now deployed, the pit lane was closed. Walkinshaw had taken over from Heyer, and the race became further in doubt as the boss waited furiously for the pit lane to re-open. While idling during the long wait for the pace car, the engine began to run on eleven cylinders after a spark plug oiled up. This was due to one of the Mahle aluminium cylinder liners loosing part of its Nicosil coating near the ring travel area after ingesting a small stone. The damage caused oil to be sucked past the oil ring and it eventually fouled the spark plug. Had the car been able to rejoin the race without any delay, the plug would probably not have fouled.

After rejoining the circuit and losing nearly a whole lap, even on eleven cylinders this XJS was the fastest car on the track, and fighting back to take a podium finish. But only first was good enough for TW at this high-profile UK meeting, so with just a few laps to go he parked the car off the circuit and walked back to the pits, telling me it had 'just stopped'. When the XJS was later retrieved, it started easily and was driven back on eleven cylinders.

There would be no picture of BMW on the top of the podium and Jaguars second and third at their home circuit, with a possible headline stating 'Jaguar Loses'. As we all knew, TW had difficulty standing on the third step of the podium if a TWR car had not won. Eating humble pie does not come easily to most people, and the chance of a negative headline exacerbating the disappointing

Little room left at this parking spot after the rain fell, but Nicholson's Jaguar managed to join in.

results could not be taken at Jaguar's home of motor racing, Silverstone, hence the decision to stop.

Kelleners/Brancatelli took first place for BMW at Silverstone with Sears/Calderari in second place for Jaguar just 34 seconds behind the winners. This act of personal pride though came at a cost. Valuable points were lost by throwing away a certain third place to add to the championship tally. Kelleners and Brancatelli closed the points gap to Walkinshaw from 56 to 27 points in just one race.

Zolder – Belgium

Zolder in Belgium two weeks later saw the usual 1-2-3 in qualifying on the tight circuit that suited the increasingly competitive Volvos and the 635i BMWs. Granberg/Lindström were only half a second off the Jaguar times in their Volvo, making good use of the homologated 'water injection' system with their turbo engines, and half a second ahead of the first BMW, driven by Stuck/Quester. Next up was the first TWR Rover of Hahne/Duez. These close gaps now left the XJS Jaguars with no clear time advantage to make up for their two stops during the race, as opposed to the one-stopping BMWs and Volvos.

BELOW
Top: The winning Granberg/Kuist Volvo 240.
Bottom: The revitalised BMW 635i of Stuck/Quester.

The Volvos had Michelin tyres, which suited one-stopping well, and would now be a real threat if they could carry enough fuel for its hungry turbo engine. The 30-litre 'water' injection tank in the passenger's foot well was proving to be a good contribution to this problem for Volvo. Henry Ford predicted that 90% ethanol with 10% water would be the fuel of the future when he was rolling out his model A Fords. How far ahead of his time was he? One-stopping cars now had the pace to be winners, and be able to continue with factory backing in 1985.

The Volvos were soon in command and by one third distance the Nicholson/Percy Jaguar was out with overheating after losing the toothed water pump belt. Rain then fell, and the pace cars were deployed to delay battle for 10 laps. After the last pit stops, the race to the end produced a win for the Volvo of Granberg/Kvist, with a greatly revitalised 635i in the hands of Stuck/Quester second and Walkinshaw/Heyer third. Calderari/Percy (Jaguar) came fourth, with the other Volvos of Lindström/Olofsson and de Dryver/Delcourt followed by Duez/Hahne in the TWR Rover.

Zolder was the penultimate race of the 1984 ETCC series, and the venue where the European Touring Car Driver's Championship was finally won by Tom Walkinshaw, with Kelleners and Brancatelli not featuring. Co-driver Hans Heyer took second place in the title because of his non-scoring round at the Nürburgring. Jaguar had at last won the Manufacturers' title and the Spa 24 Hours in its second year of official competition. It was all they could have wished for.

There was a very short celebration and sighs of relief for the team after coming second in both preceding years. The last round at Mugello in Italy was two weeks away, but in the meantime both Rover and Jaguar teams had even more work to do.

Within a few hours most of us were flying out from the Belgian airport, directly to Australia for the Bathurst 1000 race before returning for the final race at Mugello.

Mugello

After qualifying in the usual front-row positions (although with Percy ahead of Walkinshaw this time), Mugello was a non event and an anticlimax for the TWR team. First to retire was Walkinshaw with suspension problems, and then Percy had a tyre blow-out, which saw him hit the Armco barriers. The race would turn out to be a battle between Volvo and BMW, and Stuck/Ravaglia had their maiden victory, pushed by the two Volvos led by Lindström, with Kelleners/Brancatelli in fourth and Calderari/Brundle fifth in the remaining XJS.

The 1984 season had come to an end. Veteran Kelleners announced his retirement, and a weary TWR team celebrated late into the night with the local firewater.

Helmut 'The Fox' Kelleners would finally retire from active racing after the 1984 Mugello ETCC race.

It had been a great adventure for all involved in taking a car from very basic beginnings to finally win the ETCC Manufacturers' and Drivers' titles.

Qualifying was carried out on special tyres that had a life-span of just a few laps, and the Silverstone event was near the season's end each year. This disguises the huge advances made between Monza 1982, the very first race for the XJS, and the 1982 Silverstone times set towards the end of that first season.

Another point to remember is that the qualifying times set at Monza in the first race of 1983 were 10 seconds per lap faster than the XJS's first race there just 12 months earlier.

The European season had ended with the much-needed ETCC win and produced the prospect of a limited season in 1985 with TWR and Jaguar racing a prototype sports car to Group C regulations. This prospect still faced many hurdles, with issues such as what car to use – Group 44's USA design or a completely new TWR design. It was also obvious that Jaguar was short of the money needed for this step up in competition, and a period of uncertainty for TWR and Jaguar Racing began.

The Rover ETCC programme was well funded and in the process of being signed off for the next two years, including ARG Motorsport buying the services of TW as a driver for 1985 – still a closely guarded secret at the end of 1984. The expected reappearance of the XJSs in the ETCC in 1985 was not going to happen, although the public were kept in the dark over these plans.

To understand the pace of development and competitiveness in Group A over the three years, it is worth considering the reduction in times for the XJS at Silverstone during this period:

Year	Fastest qualifying lap	Fastest race lap
1982	1:42.74	1:42.72
1983	1:39.05	1:40.19
1984	1:34.68	1:38.77

But what would the two other manufacturers, Volvo and BMW do to counter their expected threat in 1985 from the XJS Jaguar? Already the problem of unregulated boost pressures for turbocharged cars in Group A was a growing reality as it was proving impossible to control this potential problem. If turbocharged cars were to be fitted with a device limiting the boost to standard pressures they would not be competitive. Manufacturers would then just provide a special run of high boost vehicles that could then be reprogrammed back to the original lower settings and sold. Unregulated boost profile could eventually spell the end for normally aspirated engines in Group A.

Although this was now the end of the XJS Jaguar era in ETCC it was not the end of the competition life for these cars. The year end would see continued expansion in Tom Walkinshaw's ambitions with another two races added to the list of events for 1984, this time on the other side of the World. Two Rovers had been entered in the new Group A division for the Bathurst 1000km race, the first time Group A cars would be seen in Australian motor sport and TW was to drive in this race as well, but in the John Goss Racing, Australian Group C, XJS Jaguar.

Two of the ETCC XJSs had now been hastily repainted in black with JPS tobacco logos ready for shipment to the Far East and the annual race around the streets of Macau.

The TWR support team in 1984. *From left:* Enzo Calderari, David Sears, Chuck Nicholson and Win Percy.

Group A – ETCC Results

1982

Destination	Drivers	Chassis No.	Position
Monza, Italy	Walkinshaw, Nicholson	001	DNF
Vallelunga, Italy	Walkinshaw, Nicholson	001	3rd
Zolder, Belgium *	Walkinshaw	001	1st
Donington, UK	Walkinshaw, Nicholson	001	DNF
Mugello, Italy	Walkinshaw, Dieudonne	001	DNF
Brno, Czechoslavakia	Walkinshaw, Nicholson	001	1st
Zeltweg, Austria	Walkinshaw	001	2nd
Nurburgring, Germany	Walkinshaw, Nicholson	001	1st
Spa, Belgium	Walkinshaw, Nicholson, Percy	002	DNF
	Dieudonne, Allam, Lovett	001	DNF
Silverstone, UK	Walkinshaw, Nicholson	002	1st
	Allam, Lovett	001	2nd
Zolder, Belgium	Walkinshaw, Nicholson	002	1st
	Allam, Dieudonne	001	2nd

1983

Destination	Drivers	Chassis No.	Position
Monza, Italy	Dieudonne, Calderari	002	DNF
	Walkinshaw, Nicholson	003	2nd
Vallelunga, Italy	Walkinshaw, Calderari	003	DNF
	Walkinshaw, Nicholson, Dieudonne	002	3rd
Donington, UK	Fitzpatrick, Calderari, Brundle	002	1st
	Walkinshaw, Nicholson	003	5th
Pergusa, Italy	Walkinshaw, Nicholson	003	1st
	Calderari	002	DNF
Mugello, Italy	Walkinshaw, Fitzpatrick	003	3rd
	Calderari, Nicholson	002	DNF
Brno, Czechoslavakia	Walkinshaw, Nicholson	003	1st
	Calderari, Dieudonne	002	6th
Zeltweg, Austria	Walkinshaw, Brundle	003	1st
	Calderari, Dieudonne	002	2nd
Nurburgring, Germany	Walkinshaw, Nicholson	003	DNF
	Calderari, Dieudonne, Walkinshaw	002	DNF
Salzburgring, Austria	Walkinshaw, Nicholson	003	1st
Spa, Belgium	Walkinshaw, Dieudonne	003	DNF
	Percy, Calderari, Brundle	002	DNF
Hockenheim, Germany*	Calderari	002	1st
	Walkinshaw	003	DNF
Silverstone, UK	Nicholson, Dieudonne	003	DNF
	Walkinshaw, Dieudonne	002	9th
Zolder, Belgium	Percy, Dieudonne, Calderari	002	DNF
	Walkinshaw, Brundle, Percy	003	8th

* Not a European Touring Car Championship race

1984

Destination	Drivers	Chassis No.	Position
Monza, Italy	Walkinshaw, Heyer	007	1st
	Brundle, Calderari	006	13th
	Percy, Nicholson	004	DNF
Vallelunga, Italy	Walkinshaw, Heyer	007	3rd
	Calderari, Nicholson	006	8th
	Percy, Schlesser	004	DNF
Donington, UK	Percy, Nicholson	004	1st
	Calderari, Schlesser	006	5th
	Walkinshaw, Heyer	007	9th
Pergusa, Italy	Calderari, Brundle	006	1st
	Walkinshaw, Heyer	007	2nd
	Percy, Nicholson	004	3rd
Brno, Czechoslavakia	Walkinshaw, Heyer	007	1st
	Percy, Nicholson	004	2nd
	Calderari, Sears	006	3rd
Osterreichring, Austria	Walkinshaw, Heyer	007	1st
	Percy, Nicholson	004	2nd
	Calderari, Sears	006	DNF
Salzburgring, Austria	Percy, Nicholson	004	1st
	Calderari, Sears	006	2nd
	Walkinshaw, Heyer	007	DNF
Nurburgring, Germany	Nicholson, Walkinshaw, Heyer	004	5th
	Heyer, Sears	007	DNF
	Percy, Calderari	006	DNF
Spa, Belgium	Walkinshaw, Heyer, Percy	004	1st
	Calderari, Sears, Pilette	006	DNF
Silverstone, UK	Calderari, Sears	006	2nd
	Walkinshaw, Heyer	007	DNF
	Percy, Nicholson	004	DNF
Zolder, Belgium	Walkinshaw, Heyer	007	3rd
	Calderari, Percy	006	4th
	Nicholson, Percy	005	DNF
Mugello, Italy	Brundle, Calderari, Sears	006	5th
	Walkinshaw, Heyer	007	DNF
	Percy, Brundle	005	DNF
Macau GP, Hong Kong*	Walkinshaw	005	1st
	Heyer	004	2nd

* Not a European Touring Car Championship race

1984 Jaguar XJ-S Group A vehicle specifications.

Wheel base. Standard 102in-2591mm.
Track. 55-58in, note offset inwards wheels reduce the track centre line by 3in.
Corner weights empty. Front 392kg – Rear 308kg. Total weight 1400kg.
Front suspension: twin wishbone, braced and spherical bearing joints.
Spring rate front. 1500-1800lb/in – 2 springs [wheel rate 1.5:1]
Rear suspension: lower wishbone & half shaft as top suspension support.
Spring rate rear. 400-450lb/in – 4 springs.
Shock absorbers. Bilstein pressurised, not external adjustable, front and rear.
Front camber 3.5° negative, Castor 4° negative, toe-out 1/8in.
Rear camber. 1.5° negative, rear toe-in 1/8in.
Front anti roll bar diameter. 1¼in. No rear roll bar.
Salisbury limited slip differential, clutch pack with shims for higher torque loading.
Ratio's used. 3.31-3.77 – Motul hypoid 80/90 oil.
Diff coolers. [2] mounted in the rear muffler wells.
Gearbox. Close ratio heavy duty Getrag 5 speed – direct 5th gear.
Clutch. AP racing high clamp load – 9.5in. single 4 puck ceramic drive plate.
Flywheel. Standard four speed manual XJS. (forged)
Front brakes. AP Racing – 14in × 1.5in with sphericone cooling vanes.
Front brake caliper. AP Racing – four 1.3/4in pistons.
Rear disc. 13in × 1in – rear brake caliper AP Racing four 1.1/16in pistons.
Brake pads. Ferodo DS11 – Motul race brake fluid
Fuel tank manufacturer. ATL – 120 litre
Roll cage material. Aluminium
Engine type V12. Bore – 90mm. Stroke – 70mm.
Inlet valve diameter 41mm. Exhaust valve diameter 35mm.
Cosworth forged pistons, 12.25:1 compression ratio.
Camshaft material. Chill cast iron or EN40 steel.
Camshaft EA3 [1983]; WJ1 [1984] – 296 degrees duration – 0.423in gross lift.
Valve train type. Duplex chain – Cosworth DFV valve springs by Schmitthelm.
Connecting rod, Special EN 24V billet forging.
Crankshaft. Standard EN 16 forging – Cross drilled big end journals, 33kg.
Main bearing cap. Standard cast iron.
Engine management. Zytek, non sequential, XK 4.2 injectors.
Ignition. HE constant energy – distributor and double coil.
Sump design. TWR recirculation wet sump, 8 litres total engine oil capacity.
Cold air intake system. Originally fed from the rear of the headlights. In 1986 the air was funneled from the top of the oil coolers into air cleaners.
Exhaust system. Special cast exhaust manifolds [4] – 2 × 2.25in secondary pipes into a 3½in tailpipe, one each bank of 6 cylinders exiting on the LH side.
Power 490bhp 1984 – 510bhp 1986 – Torque 400-410lb/ft @5000-5500rpm.
Air jacks. AP Racing [4] air jack hose connection RH side of scuttle panel.
Tyres. Dunlop crossply with Denloc double bead.
Tyre sizes front. 275/625/17 – Tyre sizes rear. 300/625/17
Wheels Speedline 17in magnesium alloy – Front 10-11in – Rear 11-11.5in wide.
Seat – Recaro racing. Seat belts – 6 point Willans.

Development of the digital engine management system

During the later part of the season and with the Jaguars dominant, the next major step forward in engine management technology was produced by Zytek. Preliminary development of the proposed V12 engine for TWR's future move to Group C had been carried out during the year in parallel with the XJS and Rover Group A engine programmes. I had used pressure coming from the development of the Lucas Micos engine management system to push Zytek to produce a digital system for TWR.

This digital system would now allow us to electronically programme and control both fuel and ignition timing. While the distributor would be retained to deliver each spark to the cylinders via the rotor, the ignition timing profile did not now rely on the distributor for triggering and providing centrifugal ignition-advance. The new system sat inside the original XJS Lucas control box, replacing the original mother board, but used existing engine looms and engine sensors. We were the first racing team in Europe to have an independent digital system with EngDiv. having access to an in-house programmable, digitally controlled total engine management system, using a double injection strategy over the 740° four stroke cycles.

The map grid was increased from 8- to 16-part throttle grid references, and retained the original 16 speed sites, making 256 separate grid plots for both fuel and ignition maps to produce a complete control system. Engine dynamometer testing produced optimised fuel and ignition numbers in each of the two separate 256 grids, a time-consuming dynamometer testing process where I was ably assisted by EngDiv.'s future Chief Engineer, 19-year-old Charlie Bamber.

Crankshaft control system

A six-slot control disc was fitted to the nose of the crankshaft and an adjustable Hall Effect trigger mounted on the front timing cover set so that the trailing edge of the No. 1 timing slot was at 10° before top-dead-centre on number one cylinder. This would produce a cranking ignition timing of 10° BTDC on start up. Each of the six timing slots would be 40° wide, and the Hall Effect trigger would communicate six engine positions over the complete 360° revolution. The Hall Effect pickup would automatically switch to be triggered by the leading edge of the blade after the engine started, allowing the computer to be signalled at 50° BTDC on each cylinder on its compression stroke. This number was selected because

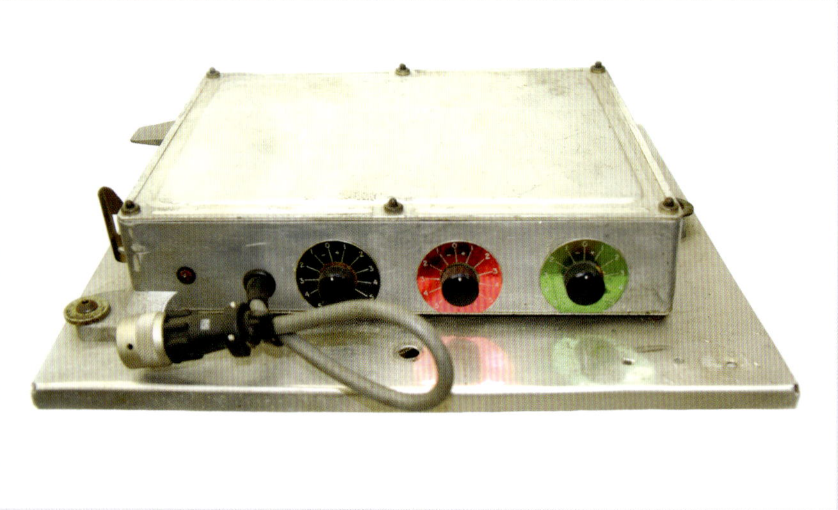

Full Digital Engine Management System with three external potentiometers for idle, ignition and fuel.

it was a point that would be ahead of any ignition timing the engine would need.

From this trigger point the software then calculated the delay needed to trigger the coil and produce the correct ignition setting from the programmed timing grid, corrected for throttle angle and engine speed.

To calculate and put this information into digital form the following was carried out.

Ignition timing programmes needed three separate calculations: Subtract the desired timing from the crankshaft timing disc slot's leading edge of 50 degrees BTDC. Multiply this figure by 256 and then divide it by 60 [the angle between each firing stroke on a V12 engine]. Type this result into the hexadecimal calculator and produce the programming hex number. This number was then typed into the EPROM programmer and written onto the set-up sheet.

All these ignition timings were at least 10° less than the 50° trigger point, allowing the computer time to calculate and deliver the various timing points to the coil.

The coil would then discharge at the correct timing position and the spark then be distributed to the cylinders using the conventional method of cap and rotor. Being able to map an ignition curve with advance/retard/advance ignition timings, something that could not be done successfully with mechanical distributor systems, would extend the power band by producing the ultimate ignition timings throughout the rpm range and have close control of detonation around the maximum torque areas. This new system necessitated special attention to the alignment of rotor and cap segment to avoid inter-cylinder crossfire, the tuning of the distributor position now having no effect on ignition timing.

With just 30° separating each segment in the 12-cylinder cap, and the need for a 15° timing window within it, correct alignment was crucial. The procedure to fix the distributor in its correct position was first to rotate the engine to 32° BTDC. Then, using a cutaway distributor cap, the rotor was aligned with the number one cylinder segment and the distributor clamped in place. The width of the rotor tip was such that a portion of it remained adjacent to a part of the cap segment throughout the ignition timing range of 30 engine degrees (15° distributor), ensuring good alignment at maximum torque ignition settings. A limited centrifugal advance ability was retained to best align the rotor and cap segment throughout the rpm range. If the rotor was set too far advanced past the cap segment, the next cylinder ahead would be subject to much lower cylinder pressures, as it would be just commencing its compression stroke. This would produce an easier path for the spark to jump the larger gap forward, and cause inter-cylinder crossfire.

Fuel map

With this digital system each grid calculation for fuel was given a fuel number between 1 and 255, and still using the 4.2 XK fuel injectors, the specified fuel number was then converted to a hexadecimal number using a hexadecimal calculator. The final calculation was typed into the EPROM programmer, and the number written onto the manual set-up sheet.

With the full programme written, the chip was programmed and then clipped into the mother board.

The Australian 'Group C' regulation John Goss Racing XJS at Bathurst.

chapter seven

Bathurst 1984

For those of us travelling to Australia the 1984 Bathurst race week had actually started at the end of the ETCC race at Zolder in Belgium, with Tom Walkinshaw taking third place and in the process winning the ETCC Manufacturers' Championship for Jaguar and confirming himself as the ETCC Drivers' Champion. The celebrations that followed were short lived for many of the team as we hastily tidied up after the race to be ready to depart for Australia.

Within 30 minutes of the end of the Zolder race we were in hire cars and after a fast drive into Brussels to a hotel close to the airport. We all crammed into two rooms to shower, change clothes, and then sprint to the terminal where an Alitalia 747 was being held for our departure. Baggage had already been checked on board and with a pre-arranged escort through customs and immigration, the TWR crew boarded the aircraft which had been held for 15 minutes, and we were on our way. The Brussels travel company responsible for the entire operation was so proud of their arrangements they made a movie of the proceedings along the way.

Before this an assault on the Bathurst 1000km race in Australia had been quietly gathering pace for some years. As early as 1981 I had invited my boss Tom Walkinshaw and his first wife, Elizabeth, to our house in the UK Cotswolds town of Long Hanborough for dinner to view a Bathurst tape which featured the Australian Ford and Holden V8s; but more importantly the Mazda RX-7 of Allan Moffat. I had spent some time living in Sydney in the early 1970s and attended the Bathurst 1000km in support of a Ford twin cam Escort entry, returning in 1978 to support this time a Mazda RX-3. I was keen to introduce the idea of TWR competing at Bathurst.

After a warm up tape that featured Aussie comedian Paul Hogan, TW was impressed with the level of competition in Australia as well as the high quality of presentation and camera work, and its potential exposure for sponsors. TWR had successfully campaigned a Group 2 Mazda RX-7 in selected races in the UK and Belgium between 1979 and 1981 and we were considering taking this car to Bathurst. After some discussion it was decided not to venture to the other side of the world as we would probably not be competitive enough and Mazda was already committed with Allan Moffat finance wise.

Fast forward to 1982 and the cover-up story for TWR's entry into the ETCC with Jaguar. To help confuse the European opposition and media attention it was said that the XJS Jaguar sitting in the workshop at Kidlington was being created for John Goss Racing, as a replacement for his existing XJS race car that had competed with limited success in the previous Bathurst 1000km races. It was from here that the beginnings of a future relationship with the famous Bathurst race began to emerge.

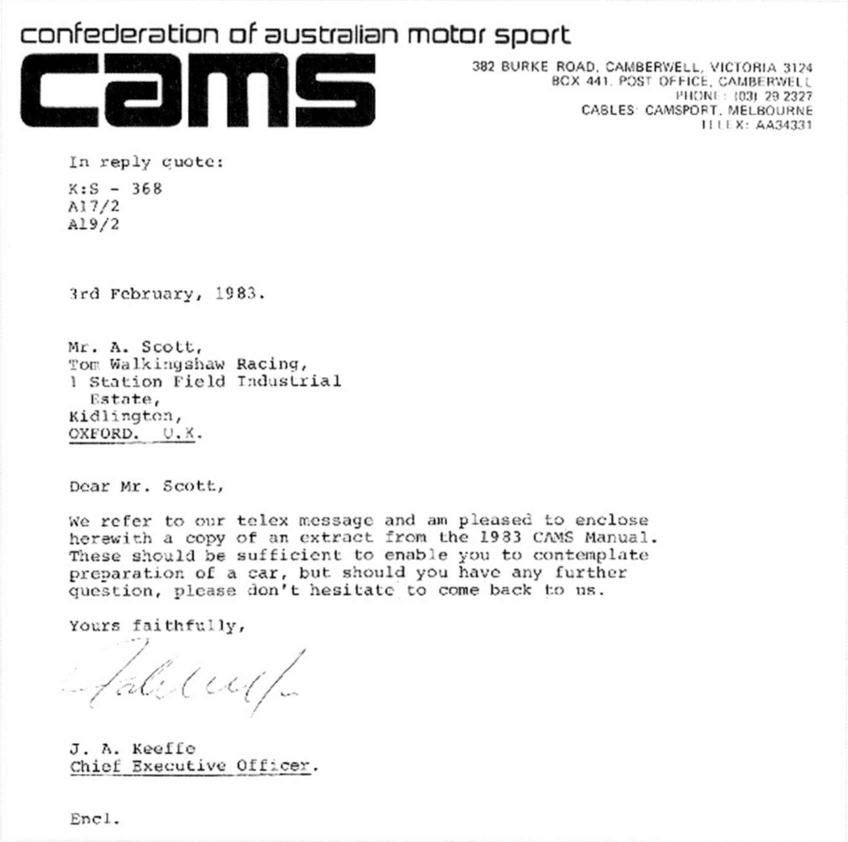

confederation of australian motor sport
CAMS

382 BURKE ROAD, CAMBERWELL, VICTORIA 3124
BOX 441, POST OFFICE, CAMBERWELL
PHONE: (03) 29 2327
CABLES: CAMSPORT, MELBOURNE
TELEX: AA34331

In reply quote:
K:S - 368
A17/2
A19/2

3rd February, 1983.

Mr. A. Scott,
Tom Walkingshaw Racing,
1 Station Field Industrial
Estate,
Kidlington,
OXFORD. U.K.

Dear Mr. Scott,

We refer to our telex message and am pleased to enclose herewith a copy of an extract from the 1983 CAMS Manual. These should be sufficient to enable you to contemplate preparation of a car, but should you have any further question, please don't hesitate to come back to us.

Yours faithfully,

J. A. Keeffe
Chief Executive Officer.

Encl.

My initial contact with CAMS in Australia.

Later, when it became apparent that Australian Group C rules would probably be switched to International Group A, I contacted the Confederation of Australian Motor Sport (CAMS) governing body for information and regulations to begin an assessment of what would be needed for TWR to compete there; but this time we had two different choices, Rover Vitesse and XJS Jaguar.

Contact with John Goss had grown via our meetings with Jaguar at Browns Lane during 1983 with various decisions being made as to how best to help him. John Goss had travelled to the UK that year for further discussions and had attended some ETCC race meetings to gain further knowledge for his Australian based team, John Goss Racing [JGR] with me in the role of liaison between teams.

A deal had then eventuated for TW to co-drive the JGR Jaguar at Bathurst in 1984 with considerable support from the TWR team. John Goss had built his XJS race car several years earlier to the local and more liberal Australian Group C regulations (not to be confused with European Group C which was for prototype sports cars), and the JGR XJS would compete in this Australian specification, not European Group A, in 1984.

TWR EngDiv. member Charlie Bamber had travelled to Sydney a few weeks before the race with parts to build an electronic fuel injection V12 engine for JGR, and to stay on as a crew member for this joint effort in the 1000km event. This engine would replace the mechanical fuel injected V12 using standard inlet manifolds but retain the tubular exhaust system used in the local regulations. With only basic workshop and engine facilities available at JGR, Charlie was taken to the edge of his comfort zone, culminating in a phone call to me at Kidlington in the week prior to us leaving for the Championship winning Zolder race. During a long conversation to update, clarify and encourage progress, he vowed he would never again complain about the facilities at TWR/EngDiv.

Now we were on our way to Australia and snatch some much needed sleep after the pressure of the last three days at Zolder to win the ETCC. Immediately after takeoff however the left wing flap would not retract fully and we were diverted to Bangkok in Thailand for repairs. Shortly after landing all passengers were processed through customs only to then have a change of plans, and we were all ushered back on board the same aircraft. It had the

The UK team with some local helpers get ready for Bathurst. The TWR regular team members from the left are: Didier Debae, Blue Dorwood, Nigel Weeks, Bob Furn, Ken Page and Grant O'Neill.

same problem, and we flew all the way to Australia with that left flap partially down accompanied by the extra wind noise.

They had arrived for their first time at Bathurst on Tuesday, after having been without any real sleep for 40 hours, and immediately started on becoming familiar with a new race car, the JGR Australia XJS, and began to fit some of the special parts that had been brought out with them. It was therefore almost two days before the exhausted team finally managed to find their beds for a well earned rest, after working through the set-up and systems for practice and qualifying the next day.

John Goss had provided the supporting equipment and race car transporter that had been kept from his earlier days as a young top Australian driver, and the 'office' at the front of the transporter where the management team now held our meetings had the added feature of sitting on a large circular padded base and being able to look at your reflection in the mirrored ceiling.

This process over the following days went reasonably well with ride heights and some other minor set up changes and all that remained to try before final qualifying was the standing start format used at Bathurst, not the more familiar rolling start of the

ETCC. Walkinshaw finally set the 8th fastest time in the JGR XJS and then 10th in the Hardies Heroes. With all the minor details to be sorted taking up all the time available, the last thing on the list, the standing start, had not been practised by TW.

With the 1984 Bathurst 1000km race ready to start on the Sunday morning let us look at what had happened from the Australian end to eventually bring this joint venture together using an Australian built XJS Jaguar supported by the TWR crew, two teams from the opposite sides of the globe.

The JGR racing project dated back to 1979 when Goss, who had won both the Australian Grand Prix driving a very potent Matich F5000 as well as the Bathurst 1000 race in a Ford Falcon, a feat not likely to be repeated, decided to build a Jaguar XJS for his next attack on Australia's premier touring car race.

John Crawford, Jaguar Rover Australia's [JRA] PR director at that time gives us an insight into the origins of Jaguar's XJS appearances on the Australian motor racing scene.

John Crawford:

We had helped John McCormack with parts for his GT racing V12 XJS built in 1979. We procured special racing items from Jaguar's Engineering Department, and gave him an introduction to the right people. He had Unipart (British Leyland) backing and John Shingleton, one of our main Australian directors, was on edge about this association after the earlier Broadspeed UK effort with Jaguar's XJC. Later when Goss came to me for support, it was obvious he was going to build his car for Bathurst whether we helped him or not, but I knew it could be embarrassing for us all if he failed.

I spoke with Ron Hancock our CEO, and we decided to give him some limited support. When he finished the car I staged two press events, one at Mallala near Adelaide, and the other at Sandown Park in Melbourne. We were also running a one-make series at Amaroo Park race track in 1980 and '81 using Triumph TR7s because while we had some difficulty selling these cars, when top drivers including Dick Johnson competed with them in televised races we sold every car we had in stock – despite the English manufacturing Triumph plant at Speke closing down. The series cost us A$75,000 to stage, but we had $55,000 in backing from Barclays Bank.

The JGR XJS under construction.

Amaroo got a new concrete pad from us, and the drivers all enjoyed an air conditioned caravan – a very professional approach there then. This positive introduction to motor sport for JGR left the door open for future promotions for us to sell cars via motorsport events.

We had sold JGR our evaluation XJS which we had in stock since 1975. It was an extremely early car, chassis 53, which we had used to launch the new model in Australia, then much later on it had been used as the prototype for a small run of lightly modified XJS cars we built to stimulate sales. Those models were limited to Australia and known as the 'Cannonball' in honour of the run in the USA which had been won by an XJS.

I also opened doors at Jaguar UK Browns Lane, so Goss then flew to England and met with Jim Randle and others.

He debuted the XJS at Bathurst in 1980 with Ron Gillard co-driving, but with not enough time to complete the project, and despite the media hype, the car was not highly modified at all.

In 1981 he brought the well-credentialed Barry (Bo) Seton in, and that year they qualified in a very improved nineteenth place. However they were out on lap seventy-three, and for us as Australia's Jaguar concessionaires we were disappointed.

By 1982 it looked as if things were improving with considerable modifications to the body work and mechanicals, and I called Jim Randle in the UK again and he got Bob Tullius to come out from the USA. Bob had run his own Group 44 team XJSs for a number

Walkinshaw and Goss with Jaguar UK marketing director Neil Johnson before the start in 1984.

of years in TransAm racing. The pair qualified in fourteenth, a big improvement, but unfortunately retired with suspension problems on lap one hundred and nineteen.

We had received a deal put forward by Ron Dickson for two of the TWR Rover Group A cars to come to Bathurst in 1984. With Australian Group C cars being in their last year of homologation and TWR's growing interest in the Bathurst event, Tom agreed to co-drive with Goss, no doubt as an introduction for a possible total TWR effort later.

It was a big deal to bring Tom Walkinshaw, the new ETCC winning driver to Bathurst to run the local Jaguar. After I had put the TWR/JGR suggestion to them, JRA heads Russell Turnham, John Shingleton and CEO Phil Hovell then put this deal together.

With these two situations at Bathurst, the Rover and Jaguar, we had quite a potentially high profile in that huge event.

UK Rover competitions head John Davenport came out to represent the Rovers as he had been involved with the co-management of the TWR cars in Europe in 1984. UK Jaguar director Neil Johnson did the same for Jaguar, so there was a lot on the line for us.

I worked hard to get all of the media coverage I could before the race started, because I understood that once the flag dropped it was a lottery. I worked on Peter Brock's principle of biting off more than I could chew, and then chewed like crazy.

How right he was.

The rear of the JGR XJS would not look like this after the start of the race.

TW had certainly enjoyed all of the printed and television coverage he was getting on arrival in Australia with promotion not seen at this level in Europe.

There were also a host of local drivers interested in being involved with any future TWR may have in Australia and here the cultural differences began to show through. TW was quite surprised telling me that 'not one of them has a good word to say about each other' with various warnings being given on who he should avoid or not trust. The British way would have been to say nothing or damn with faint praise in these situations. So after all this build up over five days we were ready for the start of the 1000km race.

The Race

While I was standing on the start line, TW had asked me through the window of the race car what I thought would be the best way to handle the standing start with the TWR/JGR/JRA XJS. This car having even larger rear tyres and corresponding grip than we ran on the TWR Group A cars. The increased rear grip and tall first gear in the Getrag gearbox was going to cause a problem for the car in this race's standing start format. It was never going to be straight forward and it was the one thing that had not been attended to on the job list.

I suggested a two-stage effort with a lower rpm start to get the car rolling, and then a harder second dip clutch start.

'Goss reckons 5000rpm, then drop the clutch,' TW had replied, and that was another option too.

John Harvey, in one of the Holden Dealer Team cars was immediately behind TW on the grid, and as I walked back past his car I indicated to John the Jaguar could have some problems getting away, not wanting him to run into the back of TW and eliminate both of us straight away. The very experienced Harvey angled his car to the inside just in case he needed to drive around the white Jaguar in front.

Group A rules allowed a single plate clutch only, and that system allowed some initial slippage when worked hard such as after a fuel and tyre stop during the ETCC races in Europe. We had no problems with the special AP Racing single plate system, but the triple plate AP clutch allowed under Australian regulations was an excellent piece of kit which allowed little or no slippage.

The XJS's race is finished at the start of the Bathurst 1000 in 1984.

As is well documented on the race video, when the green light came on the triple plate clutch simply carried out it functions too well. The grip level from the large tyres was such that the gearbox main-shaft simply twisted and snapped with no slippage from the clutch to eliminate this situation. The XJS Jaguar, with Walkinshaw sitting behind the wheel, would go no further at

Bathurst in 1984. TW's right hand was raised out of the window to indicate he had stalled, and with his eyes firmly on his interior mirror his arm would go in and out of the side window depending on the actions of the following cars trying to avoid his stationary car, and his anticipation of a violent rear impact.

All of the following competitors somehow avoided the static Jaguar, until John Tesoriero starting towards the rear of the field and by now up to high speed in his Camaro, slammed into the back of the stationary Jaguar. Peter Williamson in his Toyota Celica, then hit the Camaro leaving one badly damaged Jaguar which had travelled just 500mm in this race. Ironically, that Toyota was TWR driver Win Percy's Hughes of Beaconsfield team car which he had earlier used to win the British Touring Car Championship in 1982.

John Goss had returned to the Bathurst pit area early on Monday morning and viewed the extensively damaged rear end of his prized Jaguar. After all the build-up he was hurting both emotionally and financially, the Australian media and his peers showing little mercy to his predicaments. TW soon arrived and the two had discussions on the best way forward with Jaguar in Australia. TW wanted the Australian deal for TWR and already saw a motor racing market that impressed him. He promised John Goss a drive in any team he would bring to Bathurst in the future and Goss agreed that was the best way forward.

With another race the following weekend in Europe the team returned to Sydney for one day of R&R. The next day was spent on Sydney Harbour aboard a 45 foot cruiser, which ferried some of the team members down to Manly at the eastern end of that famous harbour. Everybody enjoyed large amounts of cooked prawns and a few beers, which was the perfect way for the very tired and disappointed team members to relax before the long flight back to Heathrow.

Peter Grace [white jacket] and I contemplate the seafood platter on Sydney Harbour.

That race distance for the Jaguar would become a record for the shortest race TWR would ever compete in – and the team had to travel from Belgium to Australia to do it. However the Rovers made up for the Jaguar's disappointment by winning the Group A section and finishing four laps ahead of the next Group A car of Denny Hulme/Prince Leopold Van Bayern in their 635i BMW. The Rover finished a very credible 12th overall against much more highly modified race cars. JRA sold Rovers for years later off the back of this Bathurst race, just reward for their support of this high profile venue using expensive European teams.

TW though, was not happy with this first outing, and was determined to come back and do it properly. He was impressed with the hype and professionalism of the race organisers and promoters as well as the leading edge technology and support from television in Australia.

There were early rumours of Group A becoming a world series such was its increasing profile and Bathurst would be one of the races on that calendar. We discussed how we would go about another attempt at winning the Bathurst 1000km race in 1985 on the long flight back to the UK.

Macau

TWR had now concluded a deal with a major tobacco company near the end of 1984 which was re-launching its JPS brand, and two of the three 1984 TWR XJSs (004 and 005) were hastily repainted in black and gold colours for shipping to Macau, close to the Chinese mainland near Hong Kong. This race was held in mid-November just three weeks after the Bathurst 1000km race. The very prestigious annual Macau GP was for Formula 3 single seater race cars and future Formula 1 World Champion Ayrton Senna had won the feature race there the year before.

The supporting races were for Group A touring cars and these end of year events, held at the end of the European racing season, were seen as just 'money races' and not taken that seriously, with no points or championships at stake.

Macau was accessed from Hong Kong by water taxi for foot passengers, or ship for the race cars as the Chinese mainland was still off limits to most high profile western promotions.

By this time the European Touring Car Championship winning race car (007) had been ear marked for enthusiast Campbell McLaren in Scotland and chassis 006 was kept in storage at Kidlington leaving 004 and 005 to be used.

The race cars had to be shipped to Macau.

TWR mechanics Graham Hodgson, Chris Tuckey and Jim McCreadie were the lead team for that mission and the main opposition for Jaguar was going to be the Marlboro-sponsored Schnitzer BMW 635is, with the remainder of the field made up of mainly local smaller engined race cars. The only other possible serious opposition might have come from New Zealander Kent Baigent in another BMW 635i.

The Jaguars were 005 for Tom Walkinshaw and his ETCC co-driver Hans Heyer 004. Following a restart after a crash involving a local driver soon after the starting flag fell, and a welcome break for the soft tyres Walkinshaw was gambling would give him the edge on this street circuit, he went on to win the 'Guia' race easily, with Hans Heyer right behind him in second. Hans Stuck in a BMW 635i was another minute behind, and was followed by Quester in the sister Schnitzer car. Baigent retired, and this win had been relatively easy for the TWR Jaguars and again confirmed to BMW that 1985 would be very difficult for them against these cars, with them having lost on a circuit that should have suited them.

And so a very eventful year had come to a close with TWR quietly switching its Jaguar team now led by Roger Silman into the new FIA Group C programme for Jaguar, to take on the seemingly impossible task of trying to defeat Porsche. Paul Davis would continue in support of the Jaguar team but his main focus had turned to managing the expanding JaguarSport business. TWR EngDiv. had started the task of providing a suitable engine for Group C and this would eventuate by mid-1985.

TW would switch to his Rover team and compete in the ETCC series over the following two years. This would become a very intensive time in Group A for all the manufacturers involved as the next chapter will show. There was still life in the Group A XJSs and special outings would continue, but not in Europe.

EngDiv. which had started in its 200 square foot workshop in 1979 would now take ove the entire 3000 square foot of the factory floor of the original Kidlington building. Some of the space vacated by Rover would be used for two new dynamometer cells to be constructed to provide much needed testing capacity for the many TWR engine programmes.

Rover would move to a new building on the Kidlington Industrial Estate.

1st and 2nd in the 1984 European Touring Car Championship for Jaguar—now, with a 1st and 2nd at Macau. TWR Jaguar Team leader Tom Walkinshaw and Hans Heyer have brought the international Group A racing season to a triumphant conclusion. They are the first Jaguar drivers to take the European title since 1963.

MACAU, Nov. 1984: Tom Walkinshaw leads Hans Heyer at 170mph along the seafront of Portugal's Chinese outpost. The Jaguars were sponsored here by John Player, the BMWs by Marlboro. Final order in the 100-mile race was: 1st Walkinshaw (Jaguar), 2nd Heyer (Jaguar), 3rd Stuck (BMW), 4th Quester (BMW).

(Photo: Chris Baker, Jaguar Service, Japan)

The winning team of Walkinshaw and Heyer.

The main contenders for Volvo in 1984.
From left: Thomas Lindström, Anders Olofsson, Ulf Granberg and Robert Kvist.

chapter eight

Volvo – BMW – Rover and the future direction of Group A

It was a time for Volvo and BMW to take stock after the 1984 season with their thinking that the XJS Jaguar would probably be even faster in1985. Few knew Jaguar's future racing plans and the occasional whispers of TWR going into Group C against the might of Porsche were greeted with the same disbelief that TWR's XJS programme had been subjected to in the early part of 1982.

We knew that the TWR Rovers would make much greater advances in performance in 1985 than both Volvo and BMW would think possible, and they had both largely discounted the Rover Vitesse as a serious contender.

BMW had begun thinking 1985 would be an interim year for them against the XJS Jaguars as they could not see the 635i being competitive, especially after being defeated at Spa in the 24 hour endurance race. BMW would move their focus to developing their new race car for 1986 based on the smaller M3 model, knowing that the Jaguar's homologation would cease at the end of 1986.

Much of this future strategy for BMW had been the result of the clever tactics that Tom Walkinshaw had taken in not revealing his hand for 1985. When eventually BMW did hear that it would be the Rovers that would be raced, a car they were confident they could beat, they had little time to switch their attention to carry out an extensive update to the 635i, still basically in 1984 specification.

Bo Wikas, Volvo's racing technical director gave these thoughts at the end of 1984

Looking at the pure race results, the performance of the 240 Turbo shook the Group A world, and the repercussions that they generated will have a huge impact on the industry. At BMW it is sit back and ask questions time, with certain consternation after a major setback for the Bavarian manufacturer who had a stronghold on the ETCC. The German management did not react very well to the Jaguar domination, which they regard as outrageously elaborate.

The new found competitiveness of the Volvo adds insult to injury, certainly with the different image between a 240 and the 635i Coupe! So in Munich, they much prefer to throw in the towel for the moment and prepare a return for the 1986 season. Maybe as Daimler-Benz thought, a light car, with a 4 valve 4-cylinder high revving engine, might be a viable alternative? The future will see. Tom Walkinshaw, responsible for the Jaguar success, believes that the future lies with the turbo engine providing that the rule makers don't put a stop to it.

But many turbo models already exist and have the required specs f or this category. At Ford, it seems that they are thinking about a solution other than the Sierra Turbo of which they had been talking for a long time: certain recent rumours led some to believe that a V8 atmospheric would be the answer [Mustang]. Could it be? No it is not possible! That would be fantastic, incredible, but not before 1986 at the earliest. In France, Danielson is ready with the Peugeot 505 Turbo, only the necessary budget is missing to mount a challenge in the ETCC. The Maserati Biturbo is a dream for many, but at what price? And then we know that Alessandro de Tomaso, owner of the brand, doesn't want to hear anything about racing following the death of Piers Courage in his Formula 1 car.

For Audi, World Rally Champion, the future of the Quattro Sport seemed vastly compromised against the upcoming Peugeot 205 Turbo 16. It would be the ideal time for them to think about the Group A and circuit racing: the Quattro 200 saloon would be an incredible weapon, and the likely success would be an incredible platform to publicise its all wheel drive concept, that would highlight the advantage on another surface.

In Japan, the Mitsubishi Starion is already homologated for Group A, not the case for the Mazda RX-7 yet, but the rotary engine also exists in a turbo version. Without forgetting Toyota or Nissan's interest in the turbo driven by the exhaust gases. Even if it is still a dream, all these possible options could generate an incredible future for the ETCC. But because of the existence of so many options, time is of the essence for Volvo if it wants to benefit from this initiative.

So far, the Swedish manufacturers rely on delivering this developed technology to a few carefully selected teams, but not with the resources required to challenge other well funded works outfits. It is insufficient to hope to achieve ultimate results, because the ETCC has become a battlefield for the biggest manufacturer. TWR represent both Jaguar and Rover, with separate teams at the circuit. As for BMW, efforts are divided between the likes of Schnitzer, Eggenberger and Juma, with logistical and human resources required. Now that there are no questions about the technical potential of Volvo in Group A, the quality of results will depend on the political sporting attitude that the manufacturers will take. In this regard it seems that they have put in place a structured effort for the coming season.

These were Volvo's thoughts at the end of 1984, showing some uncertainty as to the direction of Group A in 1985.

A test drive of the Group A race cars

Pierre Dieudonne, a former driver for TWR in both the RX-7 Mazda and the XJS Jaguar is a motor sport journalist and was also a leading Group A saloon car driver at this time. He is in the unique position of having raced or tested the main Group A contenders over the time frame that this book covers. Pierre's driving talent as well as multi lingual communication skills [he speaks French, English, Italian and a fair bit of German] in journalism made him a well sought after driver for many European teams during this period in Group A. Pierre is now based in Monaco where he fills his day by working as a consultant, writing, testing cars and staying in touch with the motor sport world. He has recently completed a large edition of the full history of Mazda motor racing, commissioned by the Mazda factory. He gives his thoughts on the Group A cars from this era.

The Volvo 240 Turbo – Pierre Dieudonne

With the Jaguar programme expanding, I was feeling less happy at TWR, which was growing into a big operation up to the point when at the end of the '83 season, I decided to stay friends with Tom and go racing elsewhere. It probably was a bit of a foolish decision of mine as, at the time, I could see nowhere to go! But it later turned into becoming a good move for me.

Towards the end of the year, I was invited to attend a test session at Zolder where a mysterious Volvo was being developed. It all seemed a very low-key operation. And it was indeed, although it evolved steadily into a bright future.

On a cold and grey day, there was that guy Thomas Lindström (then totally unknown to me) and his father, with a not so attractive Volvo 240 Turbo which they had tuned and raced in their native Sweden, supported by a few people from Volvo. It emerged that the idea was to use turbocharging to compensate for the stringent restrictions put on engine developments and tuning by the Group A technical regulations.

Moreover, the FISA rules stipulated a minimum weight scale according to the engine capacity of the cars: with an arbitrary coefficient of equivalence set at 1.4 for the turbocharged engines. Some racing-minded engineers at Volvo were anticipating interesting possibilities in terms of the power to weight ratio for turbocharged cars in Group A racing; and Volvo had a suitable model in their range!

It only took the day and a few laps in this rather odd racing car to convince me that the plan could make sense; there was good power in that relatively small and light four-cylinder 2.1 litre engine. Stepping down from a professionally run TWR-Jaguar to play around with a Volvo attended by what seemed to be a bunch of amateurs (this said with no disrespect) was not exactly a promotion. But I could see sense in that challenge.

There were a few enthusiastic guys at Volvo and they made a good job of pushing the idea forward internally. With backing from the Volvo dealers in Belgium and passion from a few people already involved in touring car racing, a team was put together by Guy Trigaux's GTM operation to run a car in the 1984 ETCC. I was to share the driving with Michel Delcourt, a jeweller from Brussels that I already knew as he was a reliable and fast touring car driver with some ETCC experience. Some other single entries would come from Sweden, the most consistent being driven by Swedish saloon specialists Thomas Lindström and Ulf Granberg, both still unknown on the international scene at that time. More famous Swedish names such as Eje Elgh and Anders Olofsson appeared during the season, as well as a second car run by GTM.

Despite some encouraging placings behind the TWR Jags and BMW 635is, it mainly was a development year, with all sorts of setbacks and reliability issues mainly linked with engine detonation. But it immediately appeared that the speed was there and the 'flying taxis' soon earned respect.

As it happened, the 1984 season ended with a clear warning when Granberg's car won at Zolder, ahead of Stuck-Quester in the factory-supported Schnitzer BMW 635i and also ahead of new ETC Champion Tom Walkinshaw in his Jag. The writing was on the wall.

And so it would be for the coming 1985 season that Volvo would take a leaf out of the TWR XJS book and have the factory build special body shells. The advantage that Volvo had over their opposition was their weight allowance of 1040kg, around 140kg lighter than the BMW and Rover. They would have similar difficulty getting down to this weight as the XJS had, and special body shells were a must. These special body shells and components included moving the chassis rails inboard by 25mm to allow for maximum tyre widths and suspension changes, just as Jaguar had done in 1982.

Detonation with increasing boost levels would be the Achilles heel of the turbocharged cars running on the mandated standard road fuel as Pierre has indicated. The homologation by Volvo of the water injection system whereby up to 30 litres of fluid could be used to help stabilize combustion, over and above its maximum fuel tank allowance, had produced a winning race car in the later part of 1984.

Separate water injection systems to stabilize combustion when using forced induction petrol based engines for fighter aircraft had been developed by both sides during WWII. The German Focke Wulf 190 had increased its engine power by 30% when later fitted with a water methanol mix injection system, while the American P-47 Thunderbolt had its 2000bhp engine increased to 2800bhp for emergencies and short maximum power periods using water injection.

This was a very radical interpretation of the rules and allowance by the FIA, and changed things for Volvo and future turbocharged cars in1985. It would also help with the Volvo's fuel consumption versus track performance and allow a strategy of one stopping in a 500km race, vital if they were to have any chance of winning.

Volvo was also in the process of producing the controversial homologation of a larger turbocharger and this would lead to a large increase in power and reliability through reduced exhaust back pressure, with higher boost capacity. This new configuration however would not be suitable for the 500 modified road vehicles on sale to the public as the homologation process was intended, and most were simply removed after inspection and replaced with the original items. Volvo would also be proved to be using non standard fuel in an effort to control engine detonation in some of their races after 1984, and this showed the increased level of intensity that was developing in Group A.

Pierre continues with his evaluation of the BMW 635 CSi

When the German manufacturer was confronted with the evidence that its 528i was not fast enough to fight off other ambitious plans in the new Group A European Touring Car Championship in 1982, it responded by homologating the Coupe 635 CSi in time for the 1983 season. Launched in 1978, the 635 CSi (E24 type) had now been revamped, with a new engine and 5-speed gearbox and front and rear aerodynamic body work. In terms of overall engineering quality and modernism, there is no question that this BMW 635i was the most suitable base model amongst all cars eligible for Group A at that time. It became immediately a consistent front-runner and would remain so until BMW turned its attention to the smaller M3 with a view to contesting the 1987 World Touring Car Championship. But such was the determination and involvement of other parties, BMWs could do no better than share honours with Jaguar, Volvo and Rover between 1983 and '86.

It was towards the end of the 1985 season that Charly Lamm of Schnitzer gave me the opportunity to test the factory-supported BMW 635 CSi raced by the likes of Roberto Ravaglia, Gerhard Berger, Hans-Joachim Stuck, Dieter Quester, Johnny Cecotto, Manfred Winkelhock, Marc Surer, Emanuele Pirro, etc. At that time, I was a Volvo driver and I had a few entertaining fights with the Schnitzer-BMWs in many ETCC races. Also, I had previous race experience of the TWR-Jaguar XJS and Rover Vitesse. It was thus obvious that this test which took place at the Salzburgring near the Schnitzer workshop, on a circuit where we had an ETCC race earlier in the year, was of mutual interest.

Located in the mountains, where winters are rigorous, the Salzburgring in Austria had a few bumps across its high speed curves and these tended to throw the car very close to the steel guard-rails running alongside the black stuff. While racing against Berger, I had noticed that my Volvo was more at ease with these bumps than his BMW. I had made the same observation through the famous and daunting section through Bosonohy village on Brno's old street course in Czechoslovakia, where the BMW's behaviour was highly spectacular. Now it was my turn to find out from the BMW's cockpit and my impression was soon confirmed. The 635 CSi was fairly stiff on its suspension and wider 11 inches tyres, but apart from being a bit harsh on bumps, its overall balance was superior to my Volvo, allowing for some acrobatic action by its star drivers. On a smooth surface, it was a better chassis altogether and a bit quicker through the corners. With a stiff body and well designed suspension, the handling was sane and predictable.

From the standard car, Schnitzer also kept the power steering and electric windows. It took a bit of time to get used to the light steering (Stuck loved it) in contrast with the high G-forces, but it certainly was a big help in long races. Most of the other BMW tuners did not fancy the power steering system and the 635 CSi was then really heavy to steer due to its suspension geometry, with plenty of castor angle.

Although the famous BMW 6-cylinders in line engine was thrilling in terms of smoothness and sound, it felt definitely down on power with 'only' 300 bhp (and the muffler fitted for the test did not help). It was still fairly early days for the Bosch Motronic management system on which BMW was doing a lot of development work in the wake of their F1 involvement and the useful rev band was a bit narrow, with a noticeable peak between 6000 and 6800rpm. With plenty of torque, the Volvo and the Jaguar were much more at ease out of the corners.

The only way for the BMW drivers to keep up was to never lose momentum through the corners and to make the most of a fairly good top speed wherever the straights were long enough (the Schnitzer BMW was clocked at 263kph on the AVUS circuit).

Late braking is another area where drivers seek to make a difference, but all these Group A cars had good powerful brakes and the limit for the braking distance was also influenced by the overall weight set by the regulations. At 1185kg, the 635 CSi was on a par with the Rover that I had tested previously, lighter than the Jaguar and heavier than the Volvo (which had narrower tyres).

At that time, BMW had two options to address the need for more power. A few weeks before my test-drive Schnitzer and BMW had run with an evolution car featuring a CSL-type rear wing and a 360bhp engine (new intake manifold, bigger valves, different camshaft, spaghetti exhaust, etc).

[**Comment: No doubt they learned from these experiments where the best results came from, looking at the Parts List on page 253.**]

In theory, BMW could have built 500 such cars and apply to the FISA for a homologation of this 635 CSi evolution. The other temptation for the motorsport people was to homologate the new M635 which featured a twin overhead camshaft 24-valves head on its 3.5 litres engine. But then it would have been necessary to build 5000 cars in one year, which was most likely to exceed the production capacity as well as the market prospects for such an exclusive car.

As it turned out, BMW opted to do no more than play with the erratum process of the homologation forms and like all Group A cars at this time stretch the rules a little. BMW was at a disadvantage engine wise with the valve size to capacity ratio of its 6-cylinder engine being less than that of the V8 Rover Vitesse. This had produced the difficulty of increasing engine power over 6800rpm for them, compared to the Rover's higher output being able to be produced at 7300rpm and able to run on to 7800rpm in 1985. BMW now made a concentrated effort to produce a basic race car under the weight limit so that lead ballast could be added to the rear bumpers to produce a weight transfer to the rear. By 1985 these cars would almost be able to achieve a 50/50 front to rear weight ratio.

Also waiting in the wings was the Ford XRTi which would have its introduction to Group A in 1985. It was becoming increasingly obvious that new special edition cars would be needed to move the goal posts further in the future, and the XRTi looked to be just the beginning of a homologation exercise for Ford in Group A with the conclusion that the future was based on turbocharging.

Some may find it difficult to change long held views on TWR's attempts in this Group A category, we were no doubt the instigators of the expanded thought process, but as time would show I think we may have even been outdone by the opposition's moves that followed 1984.

There was no doubt that Tom Walkinshaw enjoyed courting contention and some of the mystique that went with it, there are even some doubtful stories that he has encouraged.

Rover

It would not be until 1986 that TWR would be able to homologate the special 'twin plenum' induction system that EngDiv. had already developed in 1984 for the V8 Rover Vitesse competing in the French championship. Emissions approval and production of the 500 vehicles for sale would take all of 1985 to achieve. We were playing by the evolution rules, even having a special run of cross bolted V8 blocks put through production, while others were playing a game of their own.

For 1985 the Rovers would have the full digital engine management system, and higher compression ratio engines. This was not a great step forward with power increases limited to 10-15bhp. The large performance improvements for 1985 would come from an extensive upgrade of the car itself with revised suspension, special 17in Dunlop tyres and a weight reduction programme.

'You cannot be a little bit pregnant' was one of TW's favourite sayings suggesting there was no point in being just a little bit wrong, the consequences were the same for being completely wrong. You needed a good debatable argument for what you were doing even if others did not agree, and finding a 'within the rules' virgin in this series was like looking for a needle in a hay stack.

There was also no point in playing with fuel or engine capacity as there was simply no debate with those items if found out.

Both the TWR Jaguar and Rover critical induction and exhaust dimensions were never changed from the original measurements stated.

Here BMW say they made a mistake originally and used the 'Erratum' or error process to allow the exhaust manifold outlet size to be increased.

I have obtained the 'special' parts list available to BMW's few selected 635i race teams to give an idea of what was going on in Group A.

Note the 'special' cylinder head, inlet manifold, valve and even rocker arm, none of which ever went through the homologation process, and were no doubt the result of the tests that Pierre Dieudonne referred to.

This however was Group A racing and increasingly the FIA stewards and scrutineers were letting the series grow and regulate itself.

FEDERATION INTERNATIONALE
DU SPORT AUTOMOBILE

Homologation N° A-5118

Extension N° 08/04 ER

FICHE D'EXTENSION A L'HOMOLOGATION OFFICIELLE FISA
FORM OF EXTENSION TO THE OFFICIAL FISA HOMOLOGATION

☐ ET Evolution normale du type: dès le numéro de châssis / Normal evolution of the type: as from chassis number

☐ VF Variante de fourniture / Supply variant

☐ VO Variante option / Option variant

☒ ER Errata / Erratum

Homologation valable dès le / Homologation valid as from: - 1 MARS 1984 — en groupe / in group: A

Constructeur / Manufacturer: BMW — Modèle et type / Model and type: 635 csi

Page ou ext. / Page or ext.	Art. / Art.	Description / Description
12	photo J	Ø de la sortie du collecteur d'échappement :
12	photo J	Ø of the exhaust manifold exit. :
		41,5 mm

FEDERATION INTERNATIONALE DE L'AUTOMOBILE F.I.S.A.

CAMS VALID

Page 1 /

Survey: Parts not listed in the catalog

6 3 5 GR A Light-gage steel sheet

Fender fr. l.h.	2 162 805	366,--	nett.price
" fr. r.h.	"	366,--	"
Door l.h.	2 162 805	638,--	"
" r.h.	2 162 805	638,--	"
Engine hood	2 162 805	479,--	"
Trunk lid	2 162 805	414,--	"
Bumping corner	2 162 805	121,--	"
Bumper (center part)	"	262,--	"
Bumper bracket	"	35,--	"

6 3 5 GR A Glas

Set of side windows	2 162 805	614,--	"
Windscreen	"	432,--	"
Rear window	"	440,--	"

6 3 5 GR A Evolution parts

Cylinder head	2 162 805	960,--	"
Intake manifold	2 162 805	552,--	"
Bracket for intake manifold	2 162 805	11,80	"
Valve	2 162 805	35,10	"
Hub for vibration damper	2 162 805	175,50	"
Rocker arm	2 162 805	29,50	"
INJECTORS	1364 1 273270		

3 2 5 GR A Glas

Set of side windows	2 162 805	590,--	"
Windscreen	2 162 805	432,--	"
Rear window	2 162 805	380,--	"
RINGS	1125 072	92.50.	

John Crawford.

coming first in Group A, JRA's directors had warmed to the ideas being put forward of supporting a full TWR team effort to win the Bathurst 1000km Group A race outright, this time with an XJS Jaguar.

TWR's return to Australia did eventuate in September 1985 and with its three Australian Jaguar concessionaires supported XJSs it was another appearance for TWR race cars on the Bathurst track on the opposite side of the world.

Bathurst was by now unfinished business for us at TWR after the debacle in 1984 with TW's start line accident, and now TWR would return with its own three-car team, despite our very busy international schedule.

Sales director John Crawford remembers: 'We couldn't sell an XJS at the time, and had a load of new stock on the water. We were also due to finally launch an open-topped version in Australia – the Cabriolet, or XJSC as it was officially titled, and the company needed all the promotion it could get to move stock.'

JRA had finally committed to the very expensive funding of three Jaguars for Bathurst; ultimately it was JRA's Russell Turnham who had the final say in making it all happen. Crawford had attended the French Le Mans Group C race in 1984 and '85 where the USA Jaguar racing team Group 44 had brought Jaguar back to its former stamping ground with their USA IMSA specification prototype sports car. He had been impressed by Jaguar tee shirts which proclaimed: 'Jaguar Is Back' – and had then decided to use that logo for the 1985 TWR/JRA-Jaguar return to Bathurst.

1985 would be the first year of Group A racing for Australia's competitors, and clearly our XJSs had a jump on the opposition after their successes in winning the ETCC the year before. Outright victory was the only consideration, and Bathurst was increasingly valuable to Jaguar on an international level. The Seven Network's live television coverage, the best motor sport coverage in the world at the time and certainly way ahead of anything we had experienced in Europe, had a huge audience.

In Australia and New Zealand, the Bathurst weekend is sacrosanct and assured of a huge audience both at the track and on television, something only matched by Formula 1 in Europe then. The televised race was also increasingly finding its way into Asia and by now had a total viewing audience many times that of the total Australian population. Channel Seven had pioneered in-car-camera techniques, and the TWR Jaguars would get the lion's share of that coverage. TWR had previously raced without any real television exposure in the UK, and TW had enjoyed his time in front of the cameras at Bathurst the previous year.

John Crawford engaged JRA's advertising agent, Forbes McPhee Hansen, to create action packed television commercials in support, and promote 'Jaguar is back'. The race team cars were entered formally as JRA TWR Jaguar and this TWR package cost them A$740,000, a huge budget for touring car racing then. TW had managed to squeeze another £30,000 out of Jaguar Cars UK, although that was not known to JRA at the time. JRA added another A$250,000 for their advertising, making it a million dollar deal for a single race.

Despite the importance of this race it was actually a side issue for us at TWR/EngDiv. in Kidlington, with no real further development carried out on the Group A V12 engine since the beginning of 1984. Our focus in 1985 was with the huge task of turning Jaguar's V12 engine into a 700bhp fuel efficient race engine to support the new World Group C prototype sportscar programme. The XJS V12 engines for Bathurst 1985 remained as the 1984 ETCC specification with the exception of the development

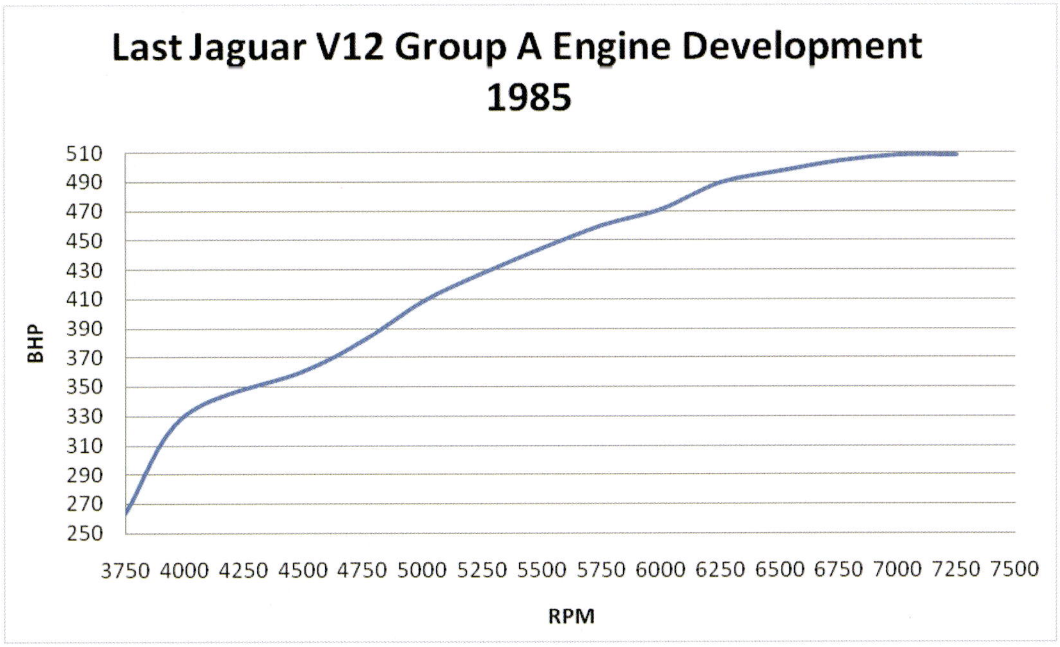

of the new in-house programmed digital engine management controls for both fuel and ignition that had now replaced the analogue system previously used in the ETCC series. By the time the cars arrived at Bathurst they were producing 500-510bhp at 7300rpm, an improvement of 10bhp, and a 5% improvement in fuel consumption.

EPROM programming for the new digital engine management system was initially laborious but at least it could now be carried out on site at EngDiv., not returned to Zytek for changes as with the earlier analogue system.

With the coaching and support of Zytek, I had been able to reprogramme the systems, and it was a large step forward for us in electronic engine technology. Vitally, EngDiv. had the ability to completely control the engine programme, and make changes at the track if needed. We also were able to fine tune the fuel mixture with the changing track altitude and air pressure on the day at Bathurst via the programmable air pressure sensor law.

Most of the improvements via the electronic controls for the Group A V12 engines at Bathurst in 1985 had come from our experiences during the year in developing dynamometer mapping techniques when fine tuning the first phase of engine development with our new fuel efficient 6222cc Group C engine now reaching 645bhp. This programme had then introduced Zytek's experimental sequential injection system that proved to be another large step forward in engine management for Group C. Phase two, a 6.5 litre V12 was already waiting to begin testing on our return to Kidlington and would produce 715bhp by early 1986.

John Crawford had previously arranged a press event during May at the prestigious Sebel Townhouse, where Formula 1 legend Stirling Moss formally revealed the three car JRA TWR Jaguar attack on 'The Mountain'. Australians had missed out on seeing an XJS in race action during 1984, but now there would be three, and with Moss headlining the revelation it had been instantly picked up by the Australian press and gained saturation coverage in the general and specialist media.

Australian drivers, Goss and Dickson, along with Rover team drivers Allam and Hahne, had tested the XJSs at Donington in the UK, Dickson being acquainted with two of the cars in June, while Goss had his workout in July. Win Percy though had tested

TWR Engineering

Engines

TWR's Engine Shop produces all the racing and road engines for the TWR Group. From the earliest days of Tom Walkinshaw's business, when Mazda RX7s with their twin rotor Wankel engines were tuned to achieve giant-killing success on the track, TWR Engineering have proven themselves more than capable of developing a powerful and reliable power unit that will live up to the rigorous and public testing it is given.

TWR Engineering have a wide experience of 4, 8 and 12-cylinder engines, from relatively mild road tuning, to full international racing specification.

There are ample facilities on site to allow building and testing of engines, with data logging of engines on test, in a new purpose-built workshop.

The 1986 6.5 litre 715bhp Group C engine in development at the end of 1985.

Ron Dickson and John Goss.

and sorted the cars thoroughly in May at three different sessions, and declared them to be faster than ever.

The cars were then shipped to Australia accompanied by a large consignment (about 30 tons) of supporting parts and equipment. TWR UK mechanic/engineer Ian 'Blue' Dorwood, originally from Hastings in New Zealand, had then gone to Australia to oversee the logistics of running the programme. The cars and Blue's skeleton team were housed at JRA's huge headquarters on the outskirts of Sydney. He was familiar with the Bathurst 1000, having worked for the Roadways and Peter Janson racing teams in the early 80s before joining TWR in the UK. Blue had arrived in Australia in early September, a full month before the race. Both he and Ron Dickson had selected some local race mechanics to assist in the preliminary setting up of the team, Grant O'Neill and the Lee brothers who had worked with Dickson and Goss previously, as well as Chris Cullen who had been independently sourced, joining TWR UK based Australian Graeme Hodgson. Their various abilities were quickly integrated with the British crew that later followed to produce an excellent mix in an amazingly short time. Everyone was totally focused on winning the race.

Things didn't start well in Australia though because several carefully constructed racks built into the shipping containers, and loaded with parts packed around each car, had broken loose during the journey collapsing onto the roof of the Walkinshaw/Percy car and damaging another to a lesser extent.

The Allam/Dickson XJS at JRA Sydney. Alva Claxton talks to Blue Dorwood.

Ken Page [back to camera] and Blue Dorwood at JRA headquarters.
Note attention to detail with colour co-ordinated trolley jack.

The roof and radio communications equipment on the 006 car were replaced before the main crew arrived, with the aerial not being fitted to the centre of the roof as the others. It was distinguished by having it fitted on the rear driver's side near the boot.

The race cars were then transported to Bathurst on the Tuesday where the XJSs were immediately fitted with various remote cameras by the Channel 7 crews. This was a good start for Jaguar and JRA with John Crawford using his connections and network of colleagues in the media to raise the team's profile. The Jaguars and TW in particular, were already becoming the stars of Bathurst that year.

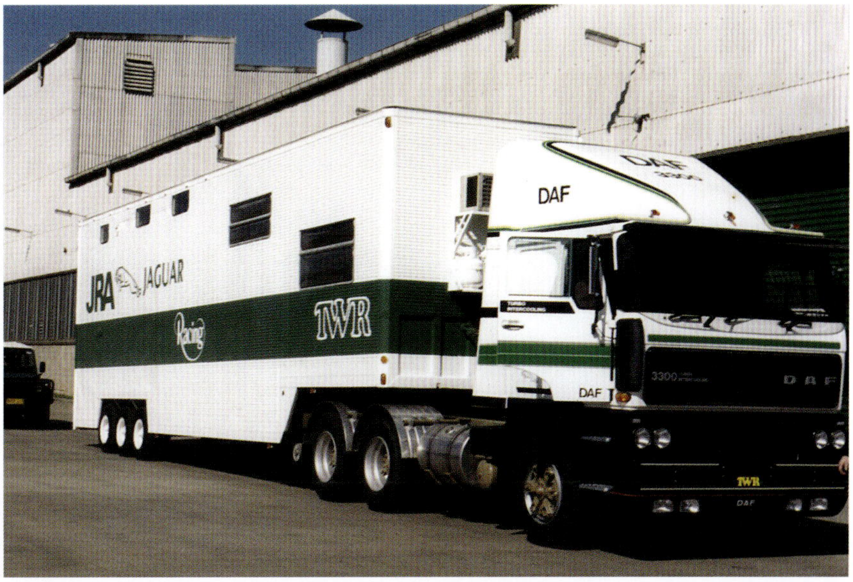

The three cars are transported to Bathurst.

The UK Jaguar support team for Bathurst would consist of the successful ETCC Rover management team which had won eight of the Group A races in Europe in 1985; however this was the first time for some of them when it came to racing an XJS Jaguar. The effort was led by Rover team manager Andy Morrison, with Eddie Hinckley as deputy and chief engineer. I had not joined, as was my priority, the Group C Jaguar team which was also

Team members: front row from left: Chris Cullen, Dave Roberts, Chris Tuckey, Grant O'Neill, Alva Claxton, Colin Marriott and Charlie Bamber, supported by the Lee brothers (Douglas, Simon, David and Christopher).

racing in Japan on the same weekend, and Group C team manager Roger Silman was only just tolerating this 'side show' as he saw it and major splitting of TWR resources. During the race weekend I would be in constant touch with the team in Japan.

The XJS drivers and management team had travelled in business class for the long journey from the UK via Singapore. My memory of the flight on the first part of the journey was time spent with Andy, Eddie and the drivers listening to TW enthuse about the new car dealerships that by now the TWR Group were becoming increasingly involved with. This was followed by a series of elimination votes being put into a hat to decide on a name for the new Porsche dealership that had just commenced trading. The winning name was 'Chariots', which was not one of my choices; all the good names had already been taken it appeared.

It should not be forgotten the incredible workload Tom Walkinshaw committed himself and this team to for the Bathurst race, literally fitting Bathurst in on the vacant weekend between two Rover races in Europe as well as TWR's main focus for Jaguar at Fuji in Japan on the same weekend. Most were impressed with what they saw with the three car TWR team but did not realise it was just part of a much larger operation. The largest factory assisted teams in Australia at the time had two race cars. TWR in 1985 had the three car XJS programme at Bathurst, a three car Rover Vitesse team in ETCC, one Rover in France and Belgium, as well as a full British rally programme, and the two sometimes three car Jaguar XJR6 Group C WSPC world series, against Porsche.

It was normally me who caused some delays as we travelled around Europe because of my New Zealand passport, but we had now arrived in Australia and the roles were reversed. I was waved straight through, but not some of the other UK team members, and Win Percy was not amused at his treatment from customs. His toilet bag was opened and checked, and even his toothpaste squeezed out. Bearing in mind Australia's early colonial history, I could only smile as I watched a slightly annoyed Win under that attention. With TW and Win featuring on television before, during, and after the race weekend, on returning to Sydney after the race for the flight home the officials were all aware of who they were; they were addressed by their first names and waved through immigration.

John Crawford had arranged a small twin engined plane for the final flight inland to rural Bathurst on the other side of the rugged Blue Mountains. The Australian Jaguar concessionaires, Jaguar Rover Australia, were looking after TW and his partner Martine, and fortunately as it would turn out, they were given alternative transport arrangements.

With many spare parts being kept for the last race at Zolder, these had now been carried to Australia in everyone's hand luggage and our small aircraft was clearly overloaded judging by the quick glances the two pilots gave each other after initial lift off when the stall warning buzzer came on. The aircraft was quickly dropped back down on the runway and continued on to increase speed to the very end of the main Mascot runway, before struggling into the air, and all the way to Bathurst as well. Virtually never in level flight at any time we all finally arrived in the early afternoon, and landed at the nearby Bathurst airport after the team had declined the offer of a quick familiarisation flight around the track. The entire team, to a man insisted on returning to Sydney after the event by car.

Most of the team were billeted on arrival at Bathurst's nearby Mitchell College, and this turned out to be one of the team's more favourable accommodations as some of the female students were still in residence.

Bathurst pits in 1985 as seen from the top of the 'Mountain' with Mitchell College in the background.

The three car team would be: Tom Walkinshaw with his 1985 Rover partner Win Percy in the lead XJS backed by Jeff Allam in a Jaguar for the first time in some years, supported by local Ron Dickson who had competed in the Bathurst race no less than eleven times. The third car was to be driven by John Goss and German ace Armin Hahne, who usually partnered his TWR Rover with Allam.

The three TWR XJSs were identified by their windscreen Jaguar signs in different colours: Chassis 004 (red 10) for Goss and Hahne; 005 (yellow 9) for Dickson and Allam; and 006 (white 8) to be driven by Walkinshaw and Percy. The cars also still carried their usual ETCC individual colour combinations in the recessed panel where the air jack pump was located.

The three XJSs in the Bathurst pits.

The beginning of the race week for JRA had been marked by a special dinner at the glamorous Sydney Regent

Hotel, and held by JRA to launch the new XJSC (Cabriolet). It had been staged on the Thursday before the race weekend, and that ensured that the most important press and television members were still in town, along with national television and all of the important 'beautiful people'. It was a great orchestration by Crawford, much better than similar exercises by Jaguar and Rover in the UK, but it added pressure to all involved with the increased expectation of a win being a foregone conclusion – not forgetting the XJS disaster the previous year, the fact that many of the team had not raced the XJSs before, and that the racing cars had not been used in competition for 12 months.

The JRA TWR Jaguar team brought a new level of professionalism in motor sport to Australia and Bathurst, and it was apparent that the sheer scale of the attack intimidated some of the established Australian teams. Two Holden Commodores would make their way to Europe to compete in the ETCC the following year, and it would prove to be a more difficult task than anticipated.

While all seemed under control from the outside a major problem had surfaced that threatened to derail the whole weekend. The cars had been sent out without any spare wheels as these were to be new stronger units manufactured by Speedline, especially for Bathurst, and air freighted from Rome. A problem with cargo space had surfaced and all the new wheels had been unloaded from the aeroplane to await another flight. Desperate phone calls to Italy had these wheels finally arrive on Friday.

In the first practice, the Jaguars were immediately the quickest cars on the circuit, Goss being the leader, Walkinshaw next quickest then Hahne. TW was increasingly confident with the situation and chose to sit out the final session on Friday. This gave the TWR crew plenty of time without any pressure to look over the car once again and fit its race engine for final qualifying the following day.

The outcome with Goss being the fastest was a talking point with some of Goss's detractors, but the plan was to make local hero Goss the fastest

Blue Dorwood at Bathurst.

before final qualifying. This would gain additional local media attention, but come Saturday TW was going to be the main man.

Hardies Heroes

'Hardies Heroes' was the title given to the top-ten shoot-out (James Hardie Industries were the main sponsor) held on Saturday afternoon before Sunday's race.

The single car run-off determined who would claim the coveted pole position, and being broadcast live was extremely important in terms of publicity. It was further reviewed on the 6 pm evening news by each of the four television networks. Peter Brock and Allan Moffat had been the main protagonists for many years in the 70s, with 'Brockie' the affable good-looking local hero, and Moffat the Canadian-born cool and rather reserved character, with his own legion of Ford devotees. Many others had also honed their own unique television styles and opposing support groups such as Dick Johnson and Allan Grice. We were now the interlopers in the Holden (GM) versus Ford battles, still a feature to this day. Now the 'Poms' from the European series had arrived to lay claim to their crown – and try to capture the partisan public's imagination; it would not be easy. I had seen though an interesting situation on the Saturday, when Peter Brock with a couple of other drivers had stopped while walking down the pit lane when TW and Win Percy had arrived. They had let TW the ETCC Champion and Win the current leader in the 1985 ETCC, walk in front of them.

Win Percy drifts through the fast left hand McPhillamy Park corner at the top of the mountain.

Australian Chris Cullen had joined the team as a mechanic just for this race and the following is his view of that experience, told to me when I met up with him again at the 2012 Melbourne Grand Prix.

> I was first contacted to be part of the Australian Bathurst support team via my connection with Les Dew who had been speaking with TWR's Blue Dorwood.
>
> After a few progress phone calls to Blue at Kidlington we first met up at Moorebank in Sydney, the JRA headquarters. This was two weeks before the Bathurst race. My boss, Paul England who had a well known race car preparation company in the '70s and '80s was very good to me, and was happy to give me the time off and have the chance of furthering my motor racing experience.
>
> On arrival at Moorebank I was impressed with the level of presentation of the three XJSs as well as all the spare parts that would ever be needed. Their organization and attention to detail with the race cars was another level above anything I had seen with any Australian team at the time. Most of TWR's racing was done away from their workshop facilities in the UK and all the equipment and spare parts were all catalogued down to the last detail, including cleaning rags. This meant the team was totally self contained and did not have to constantly look for outside assistance in unfamiliar surroundings and could operate efficiently anywhere they were racing. About 30 or 40 tons of cars and equipment all up I was told.
>
> When I first arrived I was handed a movement schedule, it was a folder with my name on it, and all the detail it involved was just amazing. It immediately made me aware of who was doing what and when and making my involvement and place in a new team that much easier.
>
> Eventually when all the team arrived, we split into three sub teams each led by a chief mechanic responsible for his three support mechanics on each car. The team management would liaise with them and they would then direct the guys on what was needed to be done. Each of the TWR mechanics had settled into their own particular part of the car's race preparation over time and this had produced a team of guys who knew what each one was doing, and if one had a problem then someone would just pitch in and help him. It was this calm confident team work that I was most impressed with, no egos and individuals with their own secret way of doing things. The English mechanics straight away pointed out how things should be done and welcomed me.

Many of the Australian teams at that time relied on unpaid helpers on the race weekends and obviously would have been better at what they did if they had been full time employees. The only guy that I knew who was on a similar level to the TWR guys in Australia was Grant O'Neill who came with John Goss and was a big help during the race weekend; he was very good.

We grew as a team in Sydney and then arrived in Bathurst and just when I thought I was coming to grips with the situation the whole atmosphere changed when the management side of TWR arrived. The team went up another gear with this whole reporting structure from Tom Walkinshaw down to the lowest guy in the team just slipping into place.

There must have been 40 or 50 in all with the JRA people and TWR had even flown out two of their own Dunlop tyre people, amazing, unheard of, even the tyres had been bought out from England. They fed us at the correct meal times, it was like a normal working day no matter what the problems might be, and supplied us with a beer at the end of the day where we were staying, but no drinking at the track at the end of the day.

I was surprised they trusted me to be the fourth guy on Tom's car. I expected to be given some minor tasks and be a gofer, but in the race I was the pit stop air jack man which was quite a responsibility. Blue had actually had me practising the pit stop and air jack work one week before in the workshop in Sydney, all part of his list of things for me to do.

I was also amazed that the engines were not to be started without the OK of the engine department. We were at Bathurst and it wasn't until Allan and his assistants Charlie and Mathew arrived and their aluminium briefcase was opened, and the electronic chips clipped into the engine management boxes that the engines were fired up. Team controlled electronics were unheard of in Australia, most of the cars still running carburettors. This starting of the engines had quite a surprising effect on me, it was as if life had now been breathed into the programme and it was alive. The sound of the V12 engines firing up just lifted my enthusiasm to another level, I had an instant adrenaline fix.

There was still a bit of scepticism about the 'pommy invaders' from many of the teams and I was called a pommy bastard a couple of times along with various jeers as I walked around in the TWR team gear. I actually felt intimidated at times from my own countrymen. I think it was also the first time the TAB was taking bets on the race and this produced a certain level of discussion and started to show some of the more negative side of some individuals. This all slowly evaporated as the team just went about its practice and qualifying runs.

It was during this time that I saw the difference in the TWR management of their racing as well. Australian teams were basically run by the driver and main mechanic who tended to be a loner and not suffer fools; also the teams not having a separate engineer in overall control of changes to the race cars. Most team managers were generally relegated to more menial organisational tasks outside of the racing effort and then lap scoring during the race, and again mainly used just on the race weekends.

With TWR at Bathurst even their top professional drivers just reported any problems to the team manager and then left it up to the management team of Tom Walkinshaw, Andy Morrison and Allan Scott, with engineering from Eddie Hinckley and Blue Dorwood, to then decide what to do. The written job list then came to us via our chief mechanic and he decided who would do the various tasks. If it was an engine problem or maintenance check, it was only carried out by Allan's team of Charlie Bamber and his assistant, Mathew Norris.

I was quite intimidated by Tom Walkinshaw and the whole business of being on TV, worried that I might make some mistake. Tom was very direct in his dealings with the team and I was surprised at how they just soaked it all up and got on with trying to keep him happy. The mechanics would just smile and out of ear-shot mimic his Scottish accent over the three main problems that they told me appeared at various times during the year; nae grip, nae grunt, nae brakes. The team just seemed to accept this and nobody became defensive or upset, it was all part of their job, no one took exception to it as we probably would have if they found nothing wrong with the race car, after Tom had thought there was.

Chris Tuckey [wheel] and Blue Dorwood.

Hardies Heroes eventually came and with the one–two results the atmosphere in the rest of the pits became more subdued, and some noticeable respect started to grow; by race day this process had continued to grow with most of the teams. Looking back I believe

> that TWR had a profound effect on the future direction of many of Australia's top racing teams. They recognised another level of business professionalism and mental attitude that had been missing, and another level higher to aim for.
>
> The rest is history and covered in this chapter on Bathurst. I would like to take this opportunity though, as I look back, to recognize the three major influences on my motor racing life. Firstly my association with Paul England, then the lessons learned from my short time with TWR that I carry to this day; and as well as the help from the best Australian race mechanic I know, Grant O'Neill.

The publicity brought to the race by live all-day and weekend-long national and international television broadcasts with all forms of media was immense, and the motivation for JRA's huge financial commitment.

Television advertisements had been running all over the country, featuring the racing XJSs, with the popular Glenn Frey song *The Heat Is On* running behind the images. This approach was very different to the non-existent BBC TV coverage for saloon cars in the UK. Although there had been brief TV coverage at the Spa 24 hour event, the first time TWR team members had actually watched their own Group A racing programme covered on commercial television had been with the Mobil Rovers at Bathurst in 1984. This high profile promotion was something which appealed to TW, who used his on camera time well. The way Australians went about their motor racing and support from the fans impressed him greatly. TW also had a good feeling for the local racing mindset and promoted himself as the team's engineering source, not just another racing driver.

Final decisions on car specification for the top ten shoot out were few as the XJS suspension and car set-ups were well established by now, leaving just the differential ratio for the long and fast Conrod Straight as the final decision. A high ratio was originally fitted to allow 7300rpm maximum on the straight, but TW soon reverted to a lower ratio because he felt the XJS was losing time up the hill, especially pulling up out of the 'Cutting' in second gear from 4300rpm on full fuel tanks. 'We will just lift off down the straight Scottie', was his typical explanation to me; something I had heard a few times before.

The option of using a lower differential ratio had evolved by 1981 with the TWR Mazda RX-7 at the Spa 24 Hours event.

We had calculated that short shifting by 300rpm and feathering the throttle on the last part of the straight did not use any more rpm per lap when using the lower differential ratio. The small improvements in lap time from better acceleration off the corners and especially in top gear on the mid part of the main straight, as well as improved overtaking situations with slower cars or changes in weather, made up for any compromise. This also left the option to improve lap times if needed by running at full rpm. Virtually all TWR's drivers could be relied on to do as asked of them in these circumstances.

Qualifying

A super-human effort by the boss on his Hardies Heroes qualifying lap, especially through the fast left hand corner after McPhilamy Park at the top of the mountain where he put his XJS up on two wheels at very high speed, bought the crowd to its feet. This effort from Walkinshaw earned him grudging acceptance from the partisan Australian crowd who live and party on the 'Mountain' for most of the week, as well as him banking the A$10,000 cheque for pole position. Many of the race fans now accepted the Scotsman as a legitimate contender. It helped cement the divide between the two racing cultures and would be the start of TWR's growing presence in Australia. His qualifying lap at Bathurst, taken from the Seven television coverage, is now a 'must see' on You Tube.

We were all hoping for a first, second and third placing on the starting grid, but when Goss had a near miss with a wayward competitor, the opportunity was lost. TW was not happy, but the team had wrapped up first, (Walkinshaw/Percy), and second (Allam/Dickson) places while the third car was to start from sixth position on the third row of the grid. It didn't look a good place to me for him to be starting from; with some of the local drivers having trouble accepting Goss's new status in the international Jaguar team.

Jeff Allam crests the mountain in practice.

The top ten placings on the grid for the race start in 1985 were:

1 Walkinshaw/Percy Jaguar
2 Dickson/Allam Jaguar
3 Johnson/Perkins Mustang
4 Grice/Cullen Commodore
5 Francevic/Bowe Volvo
6 Hahne/Goss Jaguar
7 Richards/Longhurst BMW
8 Cecotto/Ravagalia BMW
9 Harvey/Parsons Commodore
10 Brock/Oxton Commodore

Dick Johnson was third on the grid in his Ford Mustang, Allan Grice fourth in a Commodore (he would later win Bathurst for TWR in 1990 alongside Win Percy in TWR's new Holden team), and Robbie Francevic was fifth in a Mark Petch entered Volvo 240 Turbo representing New Zealand's top Group A team.

The top ten places for race day included three drivers who would at some stage in the future have an association with TWR race cars – Allan Grice, Jim Richards and Larry Perkins.

While all this was going on we were in constant touch with the Group C team at Fuji in Japan, with an increasingly agitated Group C team manager Roger Silman on the phone saying things were not going well for that particular TWR team. This was early days in the development of the V12 fuel efficient engine programme and our subcontracted machining services for new steel camshafts had not been of the quality needed, so track side changes of these items were being carried out at Fuji. This was a distraction during the Bathurst weekend for me, as I worked through these issues with TWR/EngDiv. foreman Peter Grace over the telephone to Japan.

The British team in Japan had experienced their first earthquake and a typhoon was on its way, just to add to thc alrcady tcnsc atmosphere. Those at Bathurst on the other hand were happy, however news of our good progress did not seem to make Roger any happier.

The Race

The weather at Bathurst can be changeable but race day dawned fine for what would be a six and half hour endurance event. The change from the very purposeful looking Australian Group C saloon cars with their wide flares and aerodynamic body work to the more showroom-looking Group A profiles had been expected to be something of a letdown for the public, but in fact it seemed to be an increasingly accepted change by the race fans. The international teams had arrived with the best Group A touring cars in the world and Bathurst had been lifted to a new level of intensity.

In this long race once again we would need additional fuel stops and at least one front brake pad change to that of the opposition over the 1000km race, and it was these delays that they were counting on to even up the playing field.

A large field of fifty-one cars were to start the race, including six BMW 635is and twenty-one Holden Commodores, along with one of the 1984 TWR Rovers, plus a privately entered and prepared Jaguar XJS.

The Bathurst starting procedure was unique for a Group A race in that the field circulates for one lap but then lines up for the 10 am standing start, Formula 1 style, as opposed to the rolling starts used in Europe in Group A.

It was expected that the faster Jaguars would probably drive away from the start, while John Goss would work his way through. In fact, while Walkinshaw produced a much better start than the previous year, Allam's car bogged down and was swamped into the first corner, a 90 degree left hand bend shortly after the start line. Johnson's Mustang, Francevic's Volvo, and Grice all moved passed him.

Grice's more aggressive move in his Kmart sponsored Commodore brake-tested Allam who nudged the rear of the Commodore, breaking the Jaguar's headlight glass. By the end of the first of the 162 lap race, the natural order was restored with Walkinshaw and Allam comfortably leading from Grice, Francevic, Johnson and Goss in sixth. Brock in the Holden Dealer Team Commodore, who had started in tenth, was some thirty metres behind him.

Almost immediately however the three car Jaguar team was in trouble with Allam pitting on lap three with the engine running on eleven cylinders. A quick check showed the ignition and fuel injection was not the cause, but I could hear a loud tappet noise indicating a valve gear problem. When the cylinder was isolated and spark plug removed small amounts of debris could be seen on it. That cylinder had no compression and broken glass was found in the unfiltered air intake. The cold air intake on these cars came directly from the back of the headlight housing and had never caused this type of problem in the three years we had raced the cars. Dismantling the engine later back at Kidlington showed a bent inlet valve caused by a piece of glass being trapped behind it. The valve had been jammed open and come in heavy

First laps of the race.

contact with the piston and then bent, producing a large tappet clearance and valve gear noise.

With the team down to two cars Walkinshaw picked up the pace and the Jaguars set out to put some distance between themselves and the rest of the field. This now added some further interest to the Australian teams' supporters; perhaps the Jaguars would not have it all their own way.

Walkinshaw was wired into the Seven Television Network with both a camera and microphone, and reported live that the driving standard was better than he was used to in the European Touring Car Championship: 'Where they drive like they don't have mirrors.' The locals loved it, and it was good PR if not a slight stretch of the reality; at least the 'normal' thing to now do for a slower driver was to keep a look out for his Jaguar and move out of his way.

With an hour and twenty minutes gone, Walkinshaw returned for the first scheduled pit stop and handed over to Win Percy while the team also fitted a newly fabricated wire mesh guard in the air intakes in case someone else was tempted to break another head light. That was done while fuel and tyres were attended to, along with the driver change.

Percy emerged in fifth place, with this pit stop allowing the Goss Jaguar to inherit the lead, and by recording just 255kph on Conrod Straight early in his stint, Percy was doing as asked and feathering the throttle.

Goss was in soon after, and while the previous stop had taken two minutes the swap which put Armin Hahne into the driving seat, took only one minute twenty seconds – complete with new air intake mesh guards, four tyres and fuel tank filled. This quicker stop bought the second car back into leading contention. Grant O'Neill who had earlier fabricated these mesh guards had now helped to fit them in just one and a half minutes!

With two hours gone, the Walkinshaw/Percy Jaguar was back firmly in the lead, and Hahne had passed Francevic's Volvo for second place, with Brock some 90 seconds back in fourth. The Jaguars were cruising using short shifting for lower rpm, and lifting

Walkinshaw/Percy XJS leaves Hell Corner to start the climb once again.

the throttle after the second hump on the long Conrod Straight.

Grant O'Neill races to fit the intake mesh guards.

Interestingly the TWR Group A Jaguar was at least 15kph slower than the JGR XJS we had used the year before because of its different front aerodynamic shape, with both using very similar engines. This showed just how important the design of front bodywork was on these cars when running on high speed circuits.

All the other teams banked on the Jaguars needing a longer stop for a brake pad change, plus at least one more for fuel because of their thirst, but it was mainly the brake pad changes that were seen by the opposition as the momentum breaker for the TWR team. Most teams needed at least a three minute stop and subsequent loss of a lap to carry this out in their own situations. One of the tricks here was to change the pads as early as possible, not leave them until they were worn out as many did at the time. This meant less time needed to force the caliper pistons back with wedged pads, a potentially problematic process under pressure when hurried. It would also help the opposition to think that we may need a second brake pad change.

By 12.45 pm the second stops were being planned for the leading Jaguar on lap 65, with 40% of the race distance gone and the well drilled TWR crew were ready. While other teams' personnel watched, the leading Jaguar pitted and fuel, tyres, front brake pads and driver were all changed in just forty-eight seconds. The worst fears for the other teams had come

The top contenders In the Bathurst 1000km race in 1985.

true; the TWR Jaguar team was as good, or even better, than they had been led to believe.

At 1.30 an amazing string of events saw the JPS BMW 635i challenges virtually evaporate at one corner. The talented multi Bathurst winner Jim Richards was part of the disaster, but it was certainly not of his making.

The Holden Commodore of Barry Wraith had in fact blown up at Hell Corner at the end of the Pit Straight and dropped most of its oil on the road at exactly the wrong place, just in front of the BMWs. Richards hit the oil, and immediately slid into the gravel/ sand trap narrowly avoiding an already trapped Mustang. He kept the black BMW off the Armco barriers, but was joined almost immediately by teammate George Fury in the sister JPS BMW.

Fury had also run into the driver's door of the Richards car, and both drivers were forced to try and dig their way out of the sand – believing outside help could have them disqualified. Eventually they were persuaded by the marshals they could be helped. Richards was out of the race for eleven minutes before he made his way slowly around the full length of the long track and back to the pits.

It was Hahne's turn to pit next in his XJS; his stop taking just fifty-three seconds, including changing the front disc pads, and within three laps the previous Jaguar order was restored with Brock back in third. By lap eighty-three, just over half distance, the only two cars on the same lap were the leading Jaguars.

Just after 1.30 pm Armin Hahne was out on the pit apron and ready to change over when Goss reported a problem with the driver's seat being partially broken. The driver change and refuelling were carried out, while the mechanics improvised a repair to the seat back. Zip or cable ties were placed through the holes in the top of the seat and attached to the roll cage. Hahne was sent back out to do the best he could under the circumstances, as he was offered little in the way of body support around the mountain.

The JPS BMW of Jim Richards slides into the gravel trap at Hell Corner.

Dave Roberts indicated the XJS's winning position flanked from the left by: Ken Page, Chris Tuckey, Grant O'Neill, Alva Claxton and Colin Marriott.

After the win it was a day of R and R in Sydney before the long flight home again.

Group A rules

The Group A rules were generally a very good concept, but all production based cars had some unique weakness which could cause unacceptable reliability problems. The Jaguar V12 had a steel crankshaft as standard, the BMW an inferior cast iron crankshaft prone to breakage in endurance races, similar to the problems we had with the Rover V8 engines. At the rpms and power output the XJS V12 was running, the production cast iron main bearing caps were breaking virtually every time it ran for over four hours. Trying to retain the heavy production crankshaft, which weighed 33kg, at 7800rpm, would find the weakest link, but fitment of steel main bearing caps in this situation was not allowed. Had we continued with this car we would have had to ask Jaguar to cast main cap blanks from higher grade materials and then machine and fit them to the existing engines, a time consuming and expensive process with blocks having to be remachined.

Peter Brock lost second place, we were told, with a broken timing chain. The Commodore used a production unit not designed to run with the stresses caused by the high rpms coupled to 'lash loading' associated with racing profile camshafts. TWR/EngDiv.

had encountered a similar problem with the Rover V8 engines when we first used the exact same type of timing chain. Because the later Holden engines did not have this problem again, it can be assumed they must have come to a similar conclusion as I had to fix it, by fitting a duplex chain. Rumours also did the rounds about roller rockers being used on at least one non Holden V8 engine that year because the standard cheap stamped metal units, stipulated under the rules, were unreliable and often broke.

Holden had undertaken a special selection method of rocker components and run in sets of these items for Commodore entrants to use at Bathurst. It is reasonable to assume that these were at least specially heat treated items and allowed under the rules. Holden would soon produce 500 evolution kits with roller rockers and replacement tubular exhaust manifolds which were the minimum homologation requirement, and needed to make this a competitive engine. In Bathurst 1985 specification these engines were probably producing 320-340bhp, but with the new components and another year of development, another 60-80bhp would be found.

It was pointless using any of the previously mentioned items and eliminating another potential manufacturer from competing because of a component that would constantly fail. Fine-tuning of the original specifications and rules was needed to ensure other manufacturers could reliably compete and build on this successful series. Interestingly Formula 1 for some time now has had a very experienced independent arbitrator accessible to teams and able to act in confidence with rule interpretation and changing situations.

Replacing the broken oil cooler was my call, and in hindsight I should have just bypassed it and forgotten about the oil temperature, giving TW the chance to win the race. The car was pulling 7800rpm down Conrod Straight, over 1000rpm more than any other engine in the top class that day. With the hard work needed to get up the hill, I thought the oil pressure would not have stood up to the huge temperatures which would have eventuated. We struggled to keep the temperature under control at the best of times. Although it was unknown at the time, the engine was already suffering with lower oil pressure because of two broken cast-iron main bearing caps.

During a final dinner in Bathurst, an obviously happy Armin Hahne, who couldn't help but notice the attention John Goss was getting from the media with him being claimed as the new 'King of the Mountain', came to the conclusion that since they had been driving partners, and his thinking based on the relationship of the British monarchy, he must be the 'Queen of the Mountain'. Never one for missing an opportunity, we all agreed that he was the 'Queen' of the Mountain and he should remind people of this and not let John Goss take all the glory.

On departing the Sydney airport I ventured to TW that it had been probably as good a result as any; with the local driver winning the great Australian race in a JRA TWR Jaguar. That suggestion was greeted with a look which suggested I should keep my unappreciated, antipodean's thoughts to myself. TW was still in his usual win or lose mode, and undoubtedly he never forgave me for this call. Third at Bathurst, for him, was failure. He was coming to the end of a long and successful driving career, and Bathurst was a race he desperately wanted to win.

However if you can picture John Goss sitting in the pit garage early on the Monday morning after the previous year's 1984 debacle, looking at the smashed rear section of his racing car as well as the copy of his bank statement that showed he was A$250,000 overdrawn, then in my view justice had been served.

I did not receive a win bonus for this event at year's end, despite the venture costing much less than the money raised, but was happy to accept the fact that we had won what was for me the most important XJS race, after also winning the Spa 24 Hours event in 1984.

John Crawford later said that the company sold every XJS it imported for years after on the strength of the win in 1985. It was a huge gamble for JRA Australia, but a good one for all concerned, helping to launch Tom Walkinshaw into a long and fruitful association with Holden a year later.

Importantly this historic moment in TWR and Jaguar's XJS racing history was also the last win it would achieve, despite the car racing on into early 1987. It was also virtually the end of John Goss's career despite hoping TW would pick him up for some European events.

BATHURST 1985 TWR JAGUAR

MOVEMENT SCHEDULE

SUNDAY 22nd SEPTEMBER

C.C. arrives Sydney Airport 1705 hours Ansett Flight 26. BD to collect.

Accommodation: Hunts Motel, Crossroads..

B.D.		Room 8
C.C.	S.L.	Room 19
G. O'N.		Room 11

MONDAY 23rd SEPTEMBER

R.D., B.D., C.C., S.L., G.O'N., J.W., David L., P.H., Doug L. assemble at Racing Workshop JRA Limited, Heathcote Road, Moorebank to commence final preparations to cars, spares etc. for Bathurst Race.

TUESDAY 24th SEPTEMBER

K.P., C.M., D.R., A.C., C.T. Arrive Sydney at 0830 hours on Flight AZ1768Y (Alitalia) B.D. to collect.

Accommodation: Hunts Motel, Crossroads.

K.P.		Room 11
C.M.	D.R.	Room 14
A.C.	C.T.	Room 15

SATURDAY 28th SEPTEMBER

C.B., M.N., J.McC., B.B., J.C. arrive Sydney at 0830 hours on Flight AA1768 (Alitalia) B.D. to collect and take to Bathurst in Land Rover (10 seater). R.D., K.P., C.M., A.C., C.T., D.R., C.C., G. O'N. J.W., K.P., S.L. Doug L. depart from JRA at 0900 hours for Bathurst in following vehicles.

J.W.	Race Transporter, including spares and one car
G.O'N. C.C.	JRA Leopard Bus with tyres and wheels
C.M. A.C.	Land Rover (10 seater) and car on trailer
K.P.	Land Rover (10 seater) and car on trailer
R.D.	Jaguar
D.R. C.T.	Land Rover (flat Tray) with trike and trailer
K.P.	Jaguar
S.L. Doug L.	K.P. Falcon S/W

Accommodation: From 28th September till 7th October;

B.D., K.P., A.C., C.T., C.M., D.R., G.O'N., C.C., S.L., Doug L.
K.P., J. McC., C.B., M.N., B.B., J.C. Mitchell College.

J.W.	own accommodation
R.D.	Governor MacQuarie Motel

BATHURST 1985 TWR JAGUAR

MOVEMENT SCHEDULE (Cont.)

MONDAY 30th SEPTEMBER

R. O'B arrives circuit in own transport.

Accommodation: Mitchell College till 7th October

TUESDAY 1st OCTOBER

J.D. arrives at Bathurst by 10.00 am in Jaguar

Accommodation: Governor MacQuarrie Motel till 7th October

J.G. arrives Bathurst by 10.00 am by own transport.

Accommodation: own accommodation arranged till 7th October

David L. K.L., M.D., depart Sydney p.m. for Bathurst in Blue Rover.

Accommodation: own accommodation arranged till 7th October

T.W., W.P., J.A., A.H., A.M., E.H., A.S., arrive Sydney at 0830 hours flight AZ1768 (Alitalia) Met by John Crawford (JRA) and escorted to private aircraft for flight to Bathurst Airport. Collected in 2 jaguars and Land Rover (10 seater).

Accommodation:

T.W., W.P. J.A., A.H., Governor MacQuarrie Motel till 7th October

A.M., E.H., A.S., Mitchell College till 7th October

WEDNESDAY 2nd OCTOBER

Scrutineering 2 practice sessions as per seperate sheet.

THURSDAY 3rd OCTOBER

D.N., S.D., P.H., C.P., C.L. arrive at Bathurst by 0900 in own transport

Accommodation: Till 7th October

D.N., C.P. Mitchell College

S.D., P.H., C.L. Own accommodation arrange.

Scrutineering and 2 practice sessions as per seperate sheet

FRIDAY 4th OCTOBER

Scrutineering and 2 practice sessions (Official Qualifying) as per seperate sheet.

BATHURST 1985 TWR JAGUAR

WEDNESDAY 2nd OCTOBER

Scrutineering 0800 hours to 0830 hours

Practice (untimed)
1100 hours to 1230 hours
1400 hours to 1530 hours

THURSDAY 3rd OCTOBER

Scrutineering 0800 hours to 0830 hours

Practice (untimed)
1100 hours to 1230 hours
1400 hours to 1530 hours

FRIDAY 4th OCTOBER

Scrutineering 0800 hours to 0830 hours

Practice (Timed)
1100 hours to 1230 hours

Practice (Timed nominated)
1400 hours to 1530 hours

Drivers and Managers Meeting
1000 hours ARDC Briefing Room

SATURDAY 5th OCTOBER

Scrutineering (Hardies Heroes)
0800 hours to 0830 hours

Practice HARDIES HEREOS
1000 hours to 1130 hours

Practice (untimed)
1400 hours to 1500 hours

Drivers and Managers Meeting
T.B.A. (ARDC Briefing Room)

SUNDAY 6th OCTOBER

Scrutineering 0700 hours

Race Starts 1000 hours

Drivers and Managers Meeting
0745 hours (ARDC Briefing Room)

Upon completion of race and clearing of pits, personnel may take part in the Ron Dickson Bathurst Road Show. Admission by invitation only.

BATHURST 1985 TWR JAGUAR

JAGUAR GROUP A PERSONNEL AND ROLES

Drivers:	Car 8 Tom Walkinshaw/Win Percy Car 9 Jeff Allam/Ron Dickson Car 10 Armin Hahne/John Goss
Team Manager:	Andy Morrison/Blue Dorward
Engineers:	Car 8 Eddie Hinckley Car 9 Ken Page Car 10 Blue Dorward
Mechanics:	Car 8 Colin Marriott Alva Claxton Chris Cullen Car 9 Jim McCreadie John Willard Simon Lee Car 10 Dave Roberts Chris Tuckey Grant O'Neil
Fuelers:	Douglas Lee Simon Lee David Lee
Fireman:	Rodney O'Brien
Pit Co-Ordinator:	David Lee
Storeman:	Douglas Lee
Engine Engineers:	Allan Scott (manager) Charlie Bamber Matthew Norris
Time Keepers/ Lap Scorers/General:	Janis Dickson Kerrin Lee Christopher Lee Margaret Dixon Doug Noble Ken Poulter Sonia Dickson Peter Hills Craig Poulter
Tyre Technicians:	Brian Bennett John Clifford

BATHRUST 1985 TWR JAGUAR

PIT LANE DUTIES

Race Manager:	Andy Morrison	
Radio Controller:	Eddie Hinkley	
Fuel and Oil Moniter:	Allen Scott	
Pit Crew:	Ken Page	
Wheel Changers	Jim McCreadie	Colin Marriott
	Grant O'Neil	John Willard
Air Jack	Chris Cullen	
Oil & Water	Chris Tuckey	
Fuelers:	Douglas Lee Simon Lee David Lee	
Fireman:	Rodney O'Brien	
Emergency Crew:	Blue Dorward Dave Roberts Alva Claxton Charlie Bamber	
Signaling:	Doug Noble Mathew Norris Ken Poulter	
Time Keepers/ Lap Scorers:	Eddie Hinkley Janis Dickson Kerrin Lee Christopher Lee Margaret Dixon Sonia Dickson	
Air Supply:	Peter Hills	
Fuel & Tyre Goffers:	Christopher Lee Craig Poulter	
Tyre Technicians:	Brian Bennett John Clifford	

BATHURST 1985 TWR JAGUAR

MOVEMENT SCHEDULE (cont)

SATURDAY 5th OCTOBER

Scrutineering, Hardies Heroes and Practice session as per seperate sheet

SUNDAY 6th OCTOBER

" R A C E "

MONDAY 7th OCTOBER

A.M., E.H., K.P., B.D., A.S. and others to be announced to post race scrutineering at 7.00 am.

Rest of team, excluding drivers, commence packing up for return to Sydney. Leave for Sydney upon completion of Scrutineering in same vehicles.

Accommodation: arranged in Sydney. Those people requiring same see B.D. by 1st October, 1985.

Distribution:

Tom Walkinshaw,
Win Percy,
Jeff Allam,
Armin Hahne,
Ron Dickson,
John Goss,
Andy Morrison,
Blue Dorward,
Eddie Hinckley,
Ken Page,
Alva Claxton,
Chris Tuckey,
Colin Marriott,
David Roberts,
Grant O'Neil,
John Willard
Allan Scott,
John Clifford,
Chris Cullen,
Simon Lee,
Doug Lee,
Rodney O'Brien,
Janis Dickson,
Kerrin Lee,
Christopher Lee,
Margaret Dixon,
Doug Noble,
Ken Poulter,
Sonia Dickson,
Peter Hills,
David Lee,
Jim McCreadie,
Charlie Bamber,
Mathew Norris,
Brian Bennett,
File.

Waiting for Strathmore to show up at Wellington.
From left: Armin Hahne, Win Percy and Tom Walkinshaw.

chapter ten

1986-1987: Bathurst – Fuji – New Zealand

While many hoped it might happen, most observers didn't believe Jaguar would return to Bathurst in 1986 to defend its crown. John Goss was keen, of course, and had worked hard with JRA to maximise the win, which included the introduction of the Series 3 XJ12 HE to the Australian market in 1986. JRA needed another model to run alongside the newly introduced XJ40 with its 6-cylinder engine and with the local support for V12 engines influenced to some degree by the Bathurst win in 1985, the traditional XJ12 continued on in Australia.

JRA did continue to work with TWR on the possibility of a return with the XJSs to Bathurst in 1986, and TW was keen to return and win the event. The XJSs would need all the advantages they could find after a year of Group A development for the local cars in Australia, and one car (004) was converted to left hand drive. This was in an effort to shift the weight of the driver to that side of the car to suit the Bathurst race track having mostly left hand corners from its anticlockwise direction, the opposite to tracks in Europe.

Two of the XJSs (004 and 005) were then repainted in base white with Citibank being a possible sponsor – Goss having an association with that institution.

These two cars were then tested at Silverstone by Win Percy, one in LH drive and the other original RH drive in back to back tests. Roll stiffness was increased and both cars had been subject to detailed weight reduction allowing a small increase in lead ballast in the rear. These cars were slightly faster than before at both the Silverstone and Donington tracks.

Fuji – Japan

With Bathurst a non-runner for 1986, TWR instead found backing to take the cars to Japan and New Zealand for one last foray – the XJS would then be ineligible after vehicle homologation ran out at the end of the year.

The backing came from New Zealand publisher and investment house Strathmore, and en route to New Zealand, the cars arrived in Japan for a five-hour Group A race at the Fuji International Speedway on November 9.

These races were very much a low key affair for TWR with the focus being on the increasingly competitive Silk Cut XJR-6 Group C race programme that had been in contention to beat the previously dominant Porsche teams by the end of 1986. It had been a big wake up call for Porsche with them just retaining the title by a single point, but leading to them losing their dominance the following year. The three car Rover team had been the focus of the ETCC team in 1986.

Fuji was a long way from the home base, the XJS cars had not been raced for thirteen months, and the team was unfamiliar with the track. Despite that, the TWR XJSs were the quickest cars in a straight line, and even though the Group C Silk Cut Jaguars had raced at Fuji only a few weeks before, this was not much help in setting up the Group A cars.

Walkinshaw put his car on pole, with Armin Hahne in third, split by a local 'qualifying special' Mitsubishi. When the starting flag dropped the Mitsubishi, in race trim, dropped away quickly leaving Walkinshaw and Hahne in the first two positions, with Peter Brock and Allan Grice following in Holdens.

However, after only six laps the Walkinshaw car was effectively out with a loss of engine oil pressure when its recirculation pump, just overhauled by Cosworth, seized and the drive belt broke. Johnny Cecotto in a Volvo soon disposed of

Another one for Chris Tuckey to sort out at Fuji.

both Holdens, and then claimed the lead from Armin Hahne. Fifty-year-old New Zealander and former Formula 1 World Champion Denny Hulme, having his first drive in the XJS, took over from Hahne, and he passed every car on the track except the leading Volvo.

Walkinshaw did come back for a final run in the remaining XJS, but it eventually retired with differential failure. Without ongoing development, the standard differentials were unable to reliably cope now with 510bhp from the V12 engines. These engines had been uprated in 1986 having benefited from the changes to Group A regulations. This rule change allowed unlimited valve lift in response to the fact that the turbocharged race cars' turbo boost was not and could not be regulated. It proved of little benefit engine wise to us after all the hard work already put into these engines.

The RAS Sport Volvo of Cecotto/Olofsson would go on to win from the Bob Jane sponsored Schnitzer BMW 635i of Emanuele Pirro and Charlie O'Brien and the first Japanese team of Hasemi/Tachi in a DR30 Nissan Skyline.

New Zealand

The popularity of the Jaguars in Group A allowed the XJSs to be given a one month eligibility extension to race on in New Zealand during its summer of 1987. There would be one race in Wellington, and a final outing at Pukekohe, south of the largest city, Auckland.

Unknown to most this two race series very nearly did not start for TWR with its high-flying sponsors Strathmore Group. This was all happening less than nine months before the world stock market crash and the Strathmore Group had a stratospheric share price with little to support it, as many others did at the time. I had a brief discussion with Denny Hulme at Fuji in Japan a few weeks before and he had ventured that he wondered how the Strathmore connection worked.

Andy Morrison, Blue Dorwood, Denny Hulme and Chris Tuckey at the Pukekohe test day.

Denny Hulme surveys the crashed XJS.

Their sponsorship payments were delayed and when the final demand time for the $250,000 expired 24 hours before the start of the Wellington street race, it was a magnanimous Tom Walkinshaw who decided to bite his tongue and continue on. I was later reminded by TW that it was the 'Kiwis' who had had him over, to which I apologised on behalf of New Zealand.

The Hulme/Hahne XJS was taken to Pukekohe on the 20th of January 1987 and this outing was just a test-media day, but ended with the car being crashed heavily by Armin Hahne.

There was considerable damage and Blue Dorwood's report follows.

> Mechanics were Graeme Hodgson, Chris Tuckey, and Dave Roberts with me as the hands on engineer. All four of us removed the engine and the bolt on front components. Dave, Graeme, and Chris left during the early part of the night to allow Mike Cary, our panel beater, his apprentice and myself to carry on with the reconstruction of the mangled left front corner.
>
> This all took about 24 hours of continuous work. I had tried to obtain a new replacement front sub frame as it was about the only spare that we did not have. None was available in NZ but I located one at JRA in Sydney. I phoned David Lee and placed him on standby to

The temporary pit area at the Wellington street race.

> bring it to us but the airline with the only evening flight to NZ would not allow such a very long item on the flight. I then built up a repaired front sub frame and suspension from the good range of spares we carried.
>
> The guys returned next day to refit the engine and carry out a set up. Wednesday night the car went to the paint shop and was collected the next morning and transported to Wellington in time for Friday's first practice where it ran faultlessly.

Bob Furn changing the tyres at Wellington.

Part of the identification for this chassis is the LH front air jacking point as Blue points out:

> The point about the jack tube is that I machined two in Kidlington but did not have time to drill the lightening holes before the departure of the containers to Japan. One of these undrilled tubes was welded into place along with another jack to replace the crushed set, thus giving the car an indentifying point.

On January 25 the two XJSs started in the four hour long Wellington 500, driven by Walkinshaw and Percy in 005, and Armin Hahne teamed with Denny Hulme in 004 – again ensuring more publicity with New Zealand's most successful racing driver at the wheel of a Jaguar. Rover team manager Andy Morrison was back to manage the race team.

The Wellington track was a very tight converted road layout, running through the city's down town and dock area. The Fury/Seaton DR30 Nissan Skyline had taken pole with Brock in his Commodore second and the Hahne/Hulme XJS third. The Jaguars were competitive, but they ran into tyre problems while carrying their greater weight around the tight course. With constant wheel lifting over the bumps and uneven surface the limited slip differentials were over worked and lost some of their ability to limit wheel spin, adding to the tyre degradation. Eventually this all conspired against Armin Hahne and after he had flat spotted a front tyre

Three wheeling around the streets of Wellington.

Fury is followed by the Hulme/Hahne XJS and Brock/Moffat Commodore.

he finally hit a barrier hard, and was out of the race. Win Percy retired shortly before the end of the race as a result of the differential oil cooling pump having failed, leading to the breakdown of the oil film and internal destruction.

The race was won by the HDT Commodore of Brock/Moffat with its sister car of Harvey/Lowe [NZ] in second. O'Brien/McIntyre [NZ] BMW 635i were third and the first all New Zealand team of Crosby/Wilkinson in a VK Commodore followed.

The 004 TWR XJS was now in need of an extensive chassis rebuild to the left front side of the car again and would not run at Pukekohe, and this would be the final race for 004 after its Bathurst 1000km and many other ETCC race wins.

No. 005 was the sole TWR Jaguar entered at Pukekohe on February 1, to be driven by Win Percy and Armin Hahne who had

Repairing the crashed XJS in New Zealand.

become a popular pair with the local crowds.

With just the one car available and no funding from Strathmore TW returned to the UK and Denny Hulme transferred to Larry Perkins and his Commodore for the Pukekohe race. Ironically, Perkins and Hulme went on to win the race – with the lone TWR Jaguar close behind.

The single XJS at Pukekohe.

Second is never as good as first, but you cannot always end with the best situation in motor racing, and considering the fact that any real development on the XJSs had stopped almost three years before, it was not a disgrace.

It is fitting to end a programme that I started in 1982 to have New Zealand Formula 1 Champion Denny Hulme give his feelings on driving the XJS. Denny Hulme had driven the TWR Rovers in Europe during 1986 and had become familiar with the team there. He was also my hero while serving my automotive apprenticeship in New Zealand and it was the exploits of McLaren, Hulme and Amon that motivated many of my fellow countrymen to make the journey to the UK and seek a future in motorsport.

The Pukekohe pits.

An extract from Denny's Diary in March 1987

I looked forward to driving the Strathmore Group A Jaguar XJS and working with Tom Walkinshaw's superbly organised team. As usual, TWR spared no expense, with a container load of spares, 300 tyres and two spare V12 engines. Walkinshaw's team always looked relaxed, even when the pressure was on.

Last year I would have thought the Rover Vitesse was still on as a potential winner at Wellington, but Tom reckoned there was more pace with the Jaguar. Indeed the Rover in 1987 would still be quick, but in Wellington may have made only sixth or seventh place on the grid.

The Rover is excellent on the faster circuits, but the 3.5 litre V8 lacks the horsepower of the 5 litre Holdens. In Wellington the Jaguar gearing was ideal. With so much torque and power, it is not so crucial to have gearing dead right. The twelve cylinders simply drag the car out of the corners.

We qualified on the second row of the grid with third fastest time; just over a second slower than the pole sitting Nissan Skyline. Tom Walkinshaw was set to do a flyer in the qualifying runs but his Jaguar became stuck in first gear. Tom and Win finally started from the fifth row.

Tyres of course were a major problem for the Jaguars here and we ran the hardest compound available. The amount of grunt out of the 5.3 litre V12 Jaguar engine is still staggering, and while the Wellington circuit is very tight, the green TWR cars still looked to be the cars to beat.

The Jaguar was lifting the left hand rear wheel coming through the gate around the new section onto the wharf and we were experiencing a lot of wheel spin. That powerful engine has enough zip to break away the limited slip differential and the wheel spin wore tyres out at the inner edge.

It was hot work driving the Jaguar which seems quite a bulky car. The only concession to comfort was the cooling water jacket vest and skull cap, an expensive American device. You really need to replenish the ice in the bag at each stop as this is not a refrigeration unit as some people believe. I drove the car for about an hour; my run had been relatively trouble free apart from one moment when I touched a guard rail before handing over to Armin. Just before he was due to pit at the end of his stint a tyre failed and launched him into the guardrails and we were out.

The Walkinshaw/Percy Jaguar was in good shape going into those last few minutes, looking like a strong second place-getter before the differential went. The two Jaguars suffered five punctures between them and it was inevitable that one would catch us out.

Tom flew out of New Zealand immediately following the Wellington race and it wasn't until mid-week that I heard on the radio that only one Jaguar was to run at Pukekohe.

Denny Hulme would be unaware of the funding problem with Strathmore but this was poor communication on TWR's part.

of the HE model and the first year of TWR's racing programme. The introduction of 6-cylinder and open top XJS models in 1983 boosted sales to 4317 and 5852 over the following two years and thereafter remained over 6000 until 1987, the year that the XJS stopped racing. Jaguar's motor racing investment helped to give the model an exceptionally long life (21 years) with several facelifts and derivatives, becoming the best selling Jaguar model ever at the time (115,000 units) when production finally stopped in 1996.

TWR's contribution to this success should not be underestimated with Jaguar management all agreeing that it led to a more sporting and successful image for the XJS that had suffered from having to follow the kudos associated with its predecessor, the hugely successful and more sporting E-Type, with many not identifying it as a suitable replacement.

The XJS had put Jaguar back into the international racing spotlight, and achieved all they had set out to do. The milestone of returning the Jaguar factory to private ownership in 1984 had been carried out in the very same week that the TWR team had won the Spa 24 Hours endurance race, and Tom Walkinshaw and TWR had returned some prestige to Jaguar once again through results on the race track.

TWR's bodywork enhancement kits via the JaguarSport joint venture allowed low-volume styling changes when Jaguar couldn't afford a much needed face lift version.

With the JaguarSport connection EngDiv. had carried out a complete appraisal of Jaguar's V12 5.3 litre V12 engine and these findings would eventually pave the way for the Jaguar factory to put our ideas into full production. The following is an outline of the test programme carried out in 1985 by the Race Engine Division for TWR with a view to an enhanced V12 engine for JaguarSport.

JaguarSport and EngDiv. also developed the 6 litre V12 engine that paved the way for a full-on production version to later be fitted to all V12 engined Jaguars.

EngDiv. experimental developments of Jaguar's V12 engine

I carried out an analysis of Jaguar's V12 5300cc engine in 1985 to determine what improvements could be made to enhance its performance and still pass the current emission tests for Jaguars' important American market.

I travelled to Radford, Jaguar's V12 assembly plant to observe their production methods to enable EngDiv. to produce a concept that was both compatible with the assembly line and have cost effective performance gains as well. The following tests show the process of elimination taken to produce the best outcome for the V12 engine within the guide lines of cost effectiveness, passing of emission tests, as well as improvements in performance and efficiency.

Two parameters emerged early on to guide the development path. The production process for head assembly with the automated tappet button selection for valve clearance settings dictated that this was best left as a standard process and be used with all future concepts. The added costs and emission standards when changing camshafts would reinforce this decision.

This left the design concept to focus on two directions. Direct replacement parts that could easily be substituted in the assembly process at Radford, or external components that could be substituted on new vehicles at JaguarSport's premises at Kidlington.

The initial performance focus was directed at the three speed automatic transmission that had a torque converter 'lock up' at 3500rpm under hard acceleration. Maximum torque however was above 4000rpm, with the 5.3 litre V12 not producing decent power output until 4250rpm. This tended to produce a sluggish acceleration profile between 3000 and 4250rpm as all the vehicles were fitted with automatic transmissions.

The following is a brief outline of developments to the 5300cc V12 and gives an idea of how this engine reacted to changes from the standard specification.

Note all tests were optimised for ignition timing and fueling at 3% CO.

Test 1 – This is a base line test of a standard Jaguar V12 engine running the standard car exhaust system, air boxes and paper air filters.

Test 2 – As test 1 with the special Gp.A exhaust manifold castings and larger diameter open exhausts without a silencer system.

Test 3 – As test 2 with JaguarSport larger diameter cold air intake air boxes with the standard paper air filter element removed and a K&N cotton based element substituted.

The thinking behind these modifications was that all of these elements could be added at Kidlington to a base 5.3 V12 unit fully assembled at Jaguar Radford. While these tests based mainly on

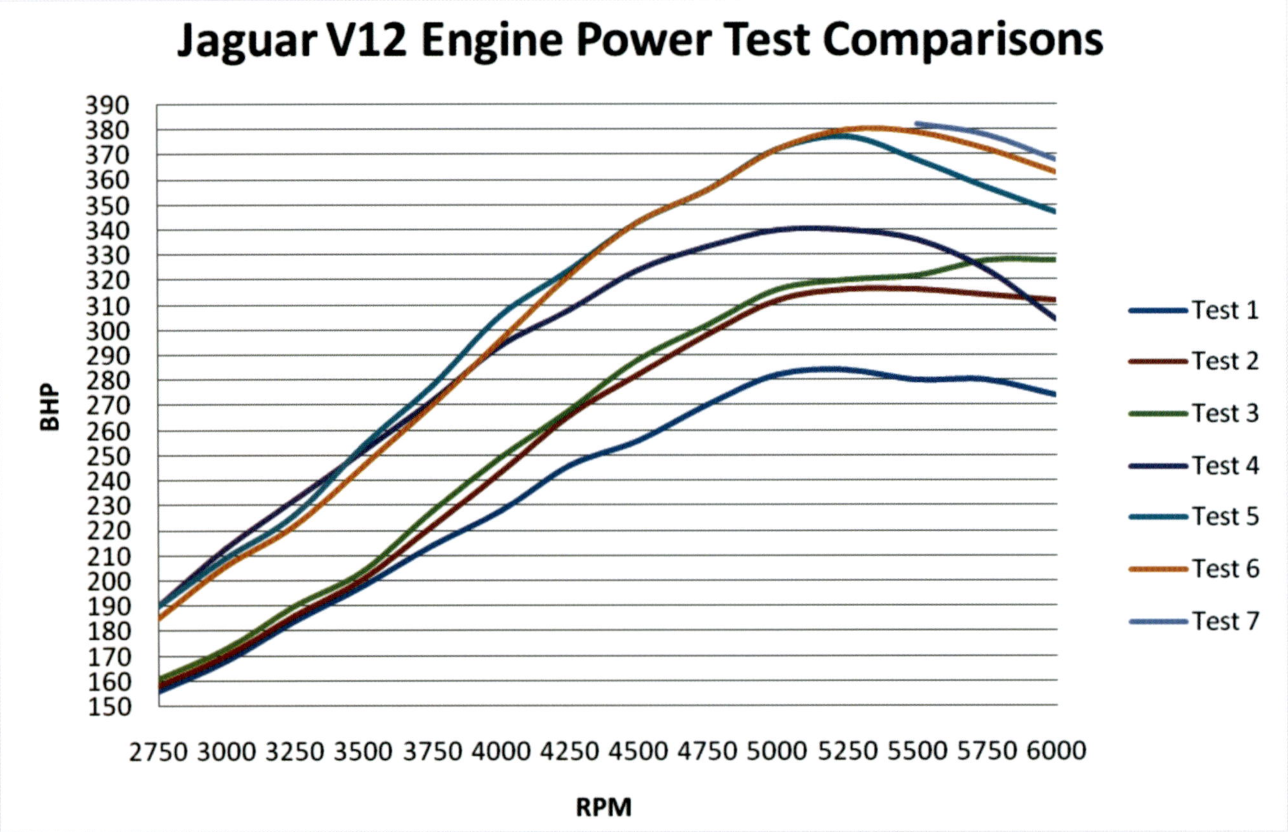

optimising the exhaust system produced an impressive power increase of over 40bhp or 23%, it was all at the top end and the effect in the 3000 to 4000rpm area was negligible until 3750rpm was reached. The test did indicate how much potential power was lost in the standard silenced exhaust system. Exhaust systems are one of the areas that modern high performance car manufacturers have developed with four tail pipe systems being now common, and from these tests it can be seen why.

Larger capacity V12

A longer stroke crankshaft of 80mm with corresponding piston design including raised gudgeon pin height to accommodate the additional stroke was produced to give an engine capacity of 6.1 litres, and tests then carried out with the aim of increasing torque in the 3000 to 4000rpm range.

The following tests on this engine were a repeat of those carried out on the standard 5.3 V12.

Test 4 – Configured as the standard 5.3 V12 [test 1] but with 80mm long stroke crankshaft fitted. It can be seen that even with all the

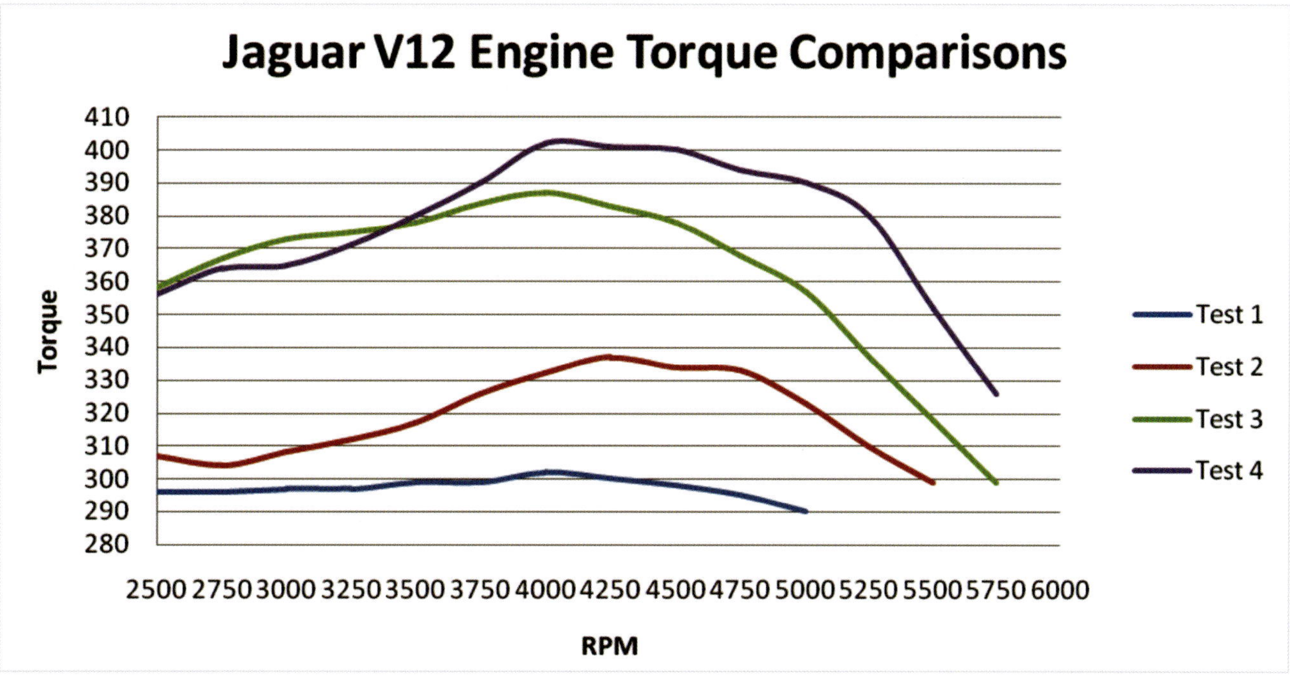

standard equipment air intakes and exhaust system this conversion has produced a huge increase in the areas we wanted to improve, with a 27% increase in torque and bhp output at 3500rpm, from a capacity increase of just 15%.

Test 5 – This had the exhaust and intake components used in test 3 on the standard 5.3 V12 fitted to the 80mm stroke engine and again showed the same trends as the 5.3 litre engine. The effects had marginal improvement in the areas we were concentrating on but came into their own once 4000rpm had been reached.

Test 6 – To see if there was any restriction when using a larger capacity V12 engine with the standard inlet manifold. The 6.1 litre engine was now fitted with the new Group A inlet manifold with its larger internal dimensions. This test produced a power reduction of 3% in the 3000 to 4000rpm area and no improvement until 5500rpm. This inlet manifold in the Group A 5.3 litre racing engine started to give good improved results at 6000rpm and proved vastly superior from 6500 to 7500rpm. These high rpms over 6000 would never be used in the road engine and this inlet manifold therefore was not what was needed for a Jaguar road car.

Test 7 – As test 6 with the standard 63mm diameter butterflies replaced with 70mm ones.

This test produced no improvements at all until over 5500rpm and then no real worthwhile gains, showing that the standard butterflies could handle additional power increases of 35% without being a restriction.

Conclusions

The best way to improve the performance of V12 Jaguar vehicles to be used with the three speed automatic transmission was to increase the capacity to six litres. This would be best facilitated by supplying a longer stroke crankshaft and piston set utilising the standard bore size of 90mm to allow assembly at Jaguar Radford. Supplying this kit of parts would mean no changes to tooling or parts handling as the crankshaft and pistons would be a direct substitute allowing 5.3 litre and 6 litre engines to be assembled on the same line.

The tests with both 5.3 and 6 litre engines demonstrated that the standard inlet manifold, valve diameters and port sizes, as well as camshaft timing actually suited a larger capacity engine more than the original 5.3 litre engine currently being used. This was a gift to the six litre conversion as it meant all standard parts could continue to be used making this proposal very cost effective.

The replacement K&N air filter showed an improvement of 1 to 2% in power throughout the range without additional fuel being added.

The 5.3 litre engine had been slightly compromised in its lower and mid ranges in efforts to produce good power outputs and continue Jaguar's status as a producer of high performance road engines. When coupled with heavier cars and three speed automatic transmissions in every day use this combination needed some refinements to produce a best compromise. The biggest single change that could have been made to help the 5.3 litre engine would have been to re-core the inlet manifold 3-4mm smaller.

The standard exhaust system showed that it started to restrain the engine after 3750rpm and produced considerable losses by 4500rpm. This exhaust restriction occurred well within the normal operating range of the road car and needed to be addressed for optimum performance. If this was not possible however it would still not compromise too much the area we wished to most improve, 3000 to 4000rpm.

The final conclusion and my recommendations to JaguarSport were that normally accepted engine improvements of camshaft design, larger valves and ported and polished heads etc were a waste of time for JaguarSport to consider. A larger capacity engine would work well with the standard components giving a much improved performance level in everyday road use. It would be able to be assembled at Radford with no changes to their assembly process and therefore be able to be mass produced at virtually no additional costs should Jaguar decide to make this larger engine a production option.

If additional horsepower was to be the aim then this would best come from changing the standard exhaust system and again use existing hardware.

The conclusions from these tests resulted in creating the JaguarSport 6 litre engine fitted with a EN 40B 78.5mm stroke crankshaft. A larger diameter exhaust and silencer system was developed as well as new air boxes with K&N air filters that took colder air from in front of the radiator instead of from the hotter engine bay as the standard engine did. Around town driving produced high under bonnet temperatures and inconsistent performance had resulted from this hot air when taken from the engine compartment.

Maximum torque was produced between 3500 and 3750rpm instead of the 4000 to 4250rpm in the 5.3 litre and this achieved the aim from the outset of making the V12 engine's power output better adapted for use with the three speed automatic transmission. The compression ration was reduced from 12.5:1 to 11.5:1 and head gaskets were changed to the type developed in the Group A race programme.

A new Zytek sequential engine management system, whose origins had been developed in the Group A race programme, was later added to better tailor the fueling and ignition needs and in the process improve fuel consumption.

At that level of development the EngDiv. JaguarSport team of two builders managed by Anthony Moss was handed over to the stand alone JaguarSport business housed on the Kidlington estate.

This 6 litre engine was later put into production at Jaguar replacing the 5.3 litre version, and all the systems and emission tests were carried out in conjunction with Zytek by TWR staff managed by Ian Brown.

A feature of the production process was the successful completion of the 400 hour test, alternating between running at full load maximum torque, and then switching to running at full load maximum power. Any failures during this process meant that the clock went back to zero and it all had to start over again.

This was completed in one continuous event stopping only for allowed checking and maintenance of external components such as dynamometer exhausts systems. These would occasionally fail through the excessive temperatures involved that would normally be lost when these systems were cooled by the air stream flowing beneath a moving vehicle.

After the test programme had been completed in 1985 a second higher performance V12 engine test was carried out with EngDiv. setting the brief for a road car V12 engine that could be hand built in limited numbers for special JaguarSport vehicles at TWR.

This engine was based around an 84mm stroke and standard bore size giving 6413cc, larger inlet valves of 44.5mm and exhaust valves of 36mm and larger Group A inlet manifold. A new camshaft set at 110 degree lobe centres and lift of 0.420in, an increase of 0.030in on the production camshaft, was supplemented with a new piston with a compression ratio of 10.5:1 This engine was tested on a short tubular exhaust manifold with 2.5in unsilenced tail pipes. Although limited development work was done in this exercise, and final settings were not optimised, it achieved its goal of a wide operating torque band from 2750rpm to 5500rpm with a variance of just 8% from its peak of 420lb/ft. Maximum power of 410bhp was achieved at 5750rpm.

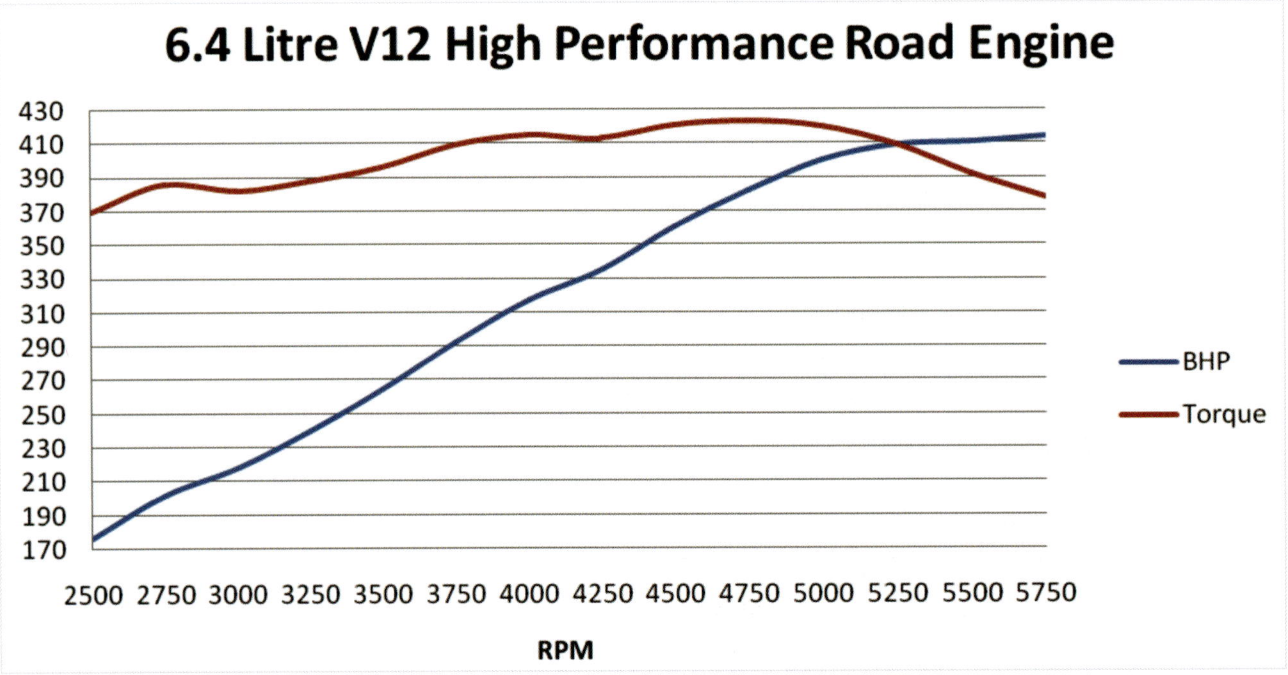

Starting in 1985 TWR's focus for Jaguar had switched from Group A to Group C with the purpose built, Tony Southgate designed, prototype Jaguar XJR6. This time the opposition would be the many Porsche teams which dominated the sports car scene, and its ultimate prize the 24 Hours of Le Mans. This race had now once again become Jaguar's goal after its many successes there in the 1950s, and TWR would be taking on a well entrenched and technically superior manufacturer with its first full season in the 1986 Championship.

Tom Walkinshaw's long and very successful racing career was also coming to an end in 1987, finally finishing in 1988 after winning more ETCC races than any other driver over the two year period 1985-86 with Win Percy in the TWR Rover V8s. He would not transfer his driving skills to the Jaguar Group C race cars, declaring this series to primarily be the domain of Formula 1 drivers. Team manager Roger Silman was keen to introduce a new Formula 1 based driver line up to his TWR Group C team.

Some of TWR's focus still remained with saloon cars however with the switch to Holden in Australia and the beginning of a long association with motor sport and special vehicle manufacturing in that country. His main personal focus in late 1986/87 was the developing relationship with General Motors Holden in Australia and to then take over the Holden Special Vehicles operation formerly run by the late Peter Brock. This involved developing the Walkinshaw 'downforce' VL Holden and included an ETCC race at the Nurburgring in 1987 as part of its development programme, and this would be his last competition outing in Europe. He also had his hands very full with the rapidly expanding TWR empire which included dealerships for Jaguars and other marques including BMW, Mazda, Nissan and Porsche.

By the mid part of 1987 with the Group C Jaguar team well into its dominance of the formerly unconquerable Porsche teams his mind was firmly on Group C racing and a Le Mans win for Jaguar. He had also made the major move in monopolising Jaguar's racing activities by clinching the American IMSA Jaguar deal to take over there from Group 44 by mid-1987.

But that's another Jaguar story and will be covered in my later book on Jaguar's V12 Group C programme.

Postscript

By the time of Tom's sad passing in 2010 after losing his battle with cancer, he had not only accumulated business interests in the UK but had concentrated his efforts in Australian motorsport with Holden's Supercar teams via his Walkinshaw Performance company. Other interests included high performance Holden road cars, plus Elfin vehicle manufacturing.

I attended his memorial service at Gloucester Cathedral in February 2011, with places limited to 850 people, and renewed long held friendships with former TWR colleagues. All volunteered that their time at TWR had been the most interesting and memorable days of their working lives.

For those who faced Tom Walkinshaw in the business of motor racing though, many felt that he sometimes pushed the boundaries a step too far, and some that formed joint relationships were surprised that these would not be for their own eventual benefit – but that should not have been a surprise. The intimidating character was often at the bottom of his business tool bag and how people reacted to that confrontation was their own prescription as to how they would be treated by him in the future. He did have respect for those that had the ability to calmly and carefully challenge him. During the time that this book is set, most teams let their results do the talking and this was true at TWR as well, but what many motor sport minded individuals who had dealings with Tom Walkinshaw missed was that he was also a salesman, selling a particular vision of his company with the aim of achieving financial success.

In early 2011 Tom's wife Martine continued with the family interest and ownership of the very successful TWR Holden V8 Supercar teams and Special Vehicles division, together with eldest son Ryan. The Walkinshaw dynasty in Australia continues, and owes it roots directly to the comeback from the Goss Jaguar failure in 1984 and to TWR's Jaguar successes in 1985 after Tom Walkinshaw's famous two wheel qualifying run over the top of the 'Mountain'.